Flexible Inkjet-Printed Antennas

Flexible Inkjet-Printed Antennas

Design, Fabrication, and Applications

edited by

Nader Shehata

JENNY STANFORD
PUBLISHING

Published by

Jenny Stanford Publishing Pte. Ltd.
101 Thomson Road
#06-01, United Square
Singapore 307591

Email: editorial@jennystanford.com
Web: www.jennystanford.com

British Library Cataloguing-in-Publication Data
A catalogue record for this book is available from the British Library.

Flexible Inkjet-Printed Antennas: Design, Fabrication, and Applications

ISBN 978-981-4968-87-4 (Hardcover)
ISBN 978-1-003-77475-4 (eBook)
DOI: 10.1201/9781003774754

Contents

Preface

Flexible antennas are robust, lightweight, conformable and higher-performance antennas that provide wireless connectivity effectively. There is an increasing trend of printing electronic circuits RF structures on flexible substrate such as typical photo-paper and PET substrate, which are promising candidates for extremely low cost substrates for printing. Such substrates have excellent dielectric characteristics and low surface profile. Through the appropriate coating, they can be suitable for direct methodologies such as conductive inkjet printing instead of the traditional metal etching techniques. In addition, paper and PET films are commercially available, cheap, and environmental friendly materials. The inkjet printing technique is fast and simple and does not require a clean room for rapid and efficient printing electronics. Inkjet printing on flexible substrate is perfect as it jets the single ink droplet from the printer's nozzles to the desired position, so there is no waste unlike the etching technique. Using silver nanoparticle inks in the inkjet-printing process ensures higher metal conductivity, compared with the bulk silver ink. This work introduces a wide experimental study on the optimum conditions for using conductive silver nanoparticles in inkjet printing of flexible nanoantennas, depending on commercial affordable printers. Different factors, including conductivity, temperature, multi-layers printing, are also investigated. Furthermore, the book includes new research on integrating the physical layer of flexible nanoantennas with smart routers, wireless sensor networks, and cyber security. This work, therefore, gives a new vision of how to integrate such flexible antennas in cyber physical systems, which supports researchers who work in this field of engineering through different applications of communications, compact sensors, and wearable electronics.

Nader Shehata

Chapter 1

Introduction to Flexible Electronics

Ahmed Mansour[a] and Nader Shehata[b]

[a]*Department of Electrical Engineering, Alexandria University, Alexandria, Egypt*
[b]*Department of Engineering Mathematics and Physics, Alexandria University, Alexandria, Egypt*

ahmed_mansour@mena.vt.edu, nader.shehata@alexu.edu.eg

1.1 Flexible Electronics

The electronics market is rapidly growing where smaller electronics need to perform multiple functions. Currently, several electronic circuits are integrated into one device to be used in different kinds of applications in normal life, such as smartphones, intelligent smart cards, energy harvesting units, foldable display screens, light-emitting diodes (LEDs), and biomedical devices [1–6], as shown in Fig. 1.1.

Multifunction electronic devices play a vital role in numerous fields. Such devices are increasingly in demand and need to be lightweight, compact size, portable, and flexible [8–11]. Therefore, the past decades have witnessed improvements in electronics, which made them more accessible for various applications.

Flexible Inkjet-Printed Antennas: Design, Fabrication, and Applications
Edited by Nader Shehata

ISBN 978-981-4968-87-4 (Hardcover), TBA (eBook)
www.jennystanford.com

DOI: 10.1201/9781003774754-1

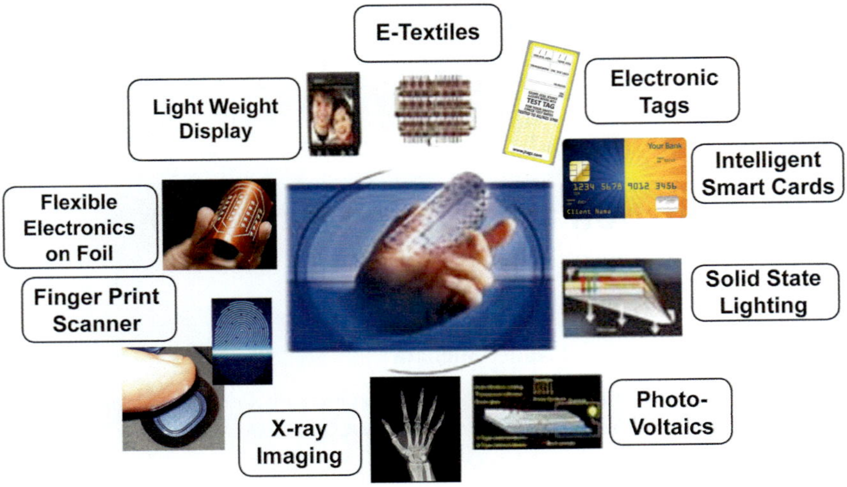

Figure 1.1 Illustration of different applications of flexible electronics in the market [7].

Flexible electronics are making their way into personal accessories and clothing, making them wearable [12, 13]. This technology raises great challenges to the substrate technology and related assembly techniques in electronic packaging for high density, small feature size, and high performance. This will open a new era of smart and intelligent flexible device technology [14–17].

Thus, the formation of highly reliable, foldable, and flexible electronic devices has become an important factor in increasing the functionality of future devices. The substrate material has sufficient area to carry all the components and can be folded into any shape, which improves the characteristics and reduces the overall device footprint. This enhances portability and functionality and makes their substrate suitable for different wearable applications [11, 13, 15]. Currently, the major interest is developing small-size electronic devices and RF structures that are discreet and flexible with extremely low cost of fabrication. Such devices will be used in mass consumer products based on accurate and low-cost printing techniques such as screen and inkjet printing. Moreover, printing on renewable and environmentally friendly materials, such as paper, wood, and clothes, is an efficient solution and provides a low-cost fabrication process [18–20].

A new trend in printing electronic circuits and RF structures on flexible substrates is to have lightweight, conformable, and higher-performance substrates. Such substrates are typical photo paper and polyethylene terephthalate (PET) films, which are promising candidates for extremely low-cost substrates for printing. These substrates have excellent dielectric characteristics and low surface profiles [21–25]. Through the appropriate coating, they would be suitable for direct-write methodologies such as conductive inkjet printing, instead of traditional metal etching techniques. In addition, paper and PET films are commercially available, cheap, and environmentally friendly materials [26–30]. Regarding the printing technique, the inkjet technology is the most suitable for different applications and it provides a fast, simple, and low-cost technique, as it does not require a clean room for rapid and efficient printing electronics. Inkjet printing on a flexible substrate is perfect, as it jets the single ink droplet from the printer's nozzles to the desired position, so there is no waste unlike the etching technique [31–33]. Nanoelectronics is continually playing a vital role in improving the capability of electronic products [16, 17].

This technology made the devices very light and easy to carry, and at the same time, it has enhanced their performance and reduced the power requirements. Using nanoparticle inks such as silver, the inkjet printing process ensures higher metal conductivity, as it has been selected to implement the circuits and improve the conductivity depending on their higher surface-to-volume ratio with high accuracy and to enhance thermal and electrical conductivity, compared to the bulk silver ink [34]. The following section presents a brief history of nanotechnology developments during the past decades, showing how nanotechnology is involved in current entire life several applications.

1.2 Integration Between Nanotechnology and Electronics

Nanotechnology was first discussed by Richard Feynman in 1959 when he talked about the existence of plenty of room at the bottom, and he described the possibility of synthesis via direct manipulation of atoms [20, 35]. Later, in 1974, the term "nanotechnology" was first defined by Norio Taniguchi of Tokyo Science University in his paper, in which he stated: "Nanotechnology mainly consists of the processing of separation, consolidation, and deformation of materials by one atom or one molecule" [36].

From this moment, the era of nanotechnology rose to the realm and researchers defined it as the study of the controlling of matter on an atomic and molecular scale [37]. Generally, nanotechnology deals with structures in the nanometer size. It involves developing materials or devices within that size. Nanotechnology is very diverse, ranging from extensions of conventional device physics to completely new approaches based upon molecular self-assembly, and from developing materials with dimensions on the nanoscale to investigating whether we can directly control matter on the atomic scale in several fields. One of the vital fields of nanotechnology is nanoelectronics, which plays a significant role in the fields of engineering and electronics [38–40].

At the beginning of the 1980s, IBM invented the scanning tunneling microscope (STM) as an instrument for measuring and identifying structures at the nanoscale, with a lateral resolution of 0.1 nm and depth resolution of 0.01 nm [41]. Thus, individual atoms within materials could be routinely imaged and manipulated. Later in the 1980s, Drexel published his research on molecular machinery or molecular nanotechnology (MNT), a technology that is based on the ability to build structures to complex atomic specifications through mechano-synthesis [42, 43]. Based on Richard Feynman's vision of miniature factories using nanomachines to build complex products, this advanced form of nanotechnology or molecular manufacturing would make use of positional controlled mechano-synthesis guided by molecular machine systems, which combined biophysics, chemistry, and molecular machinery. In the future, jets, rockets, cars, or even chairs will be able to be fabricated, as remarkably lightweight, strong, and inexpensive [43].

Molecular surgical tools, guided by molecular computers and injected into the bloodstream, could find and destroy cancer cells or invading bacteria, unclog arteries, or provide oxygen when the circulation is impaired [44].

Finding a new, highly stable form of a pure element is rare in the world of chemistry, and for this reason alone, the discovery of buckyballs was noteworthy. But it also opened a whole new field of chemical study with practical applications that scientists are only beginning to uncover. Buckyballs were the first of a whole class of hollow, closed-shell carbon macromolecules that came to be known as fullerenes. They have become the subject of intense research, both for their unique chemistry and technological applications, especially in the materials science, electronics, and nanotechnology areas [45].

Perhaps the most significant fullerenes to emerge since the buckyballs are carbon nanotubes, or "buckytubes", discovered in 1991 by Lijima Sumio of NEC Corporation in Tsukuba, Japan [46]. A buckytube is a carbon sheet rolled into the shape of a tube or cylinder, capped with round domelike ends and with a diameter of approximately a single nanometer, or one-billionth of a meter, but extremely long by comparison. These long, thin tubes opened further research by Smalley and others in fullerenes and their applications. Since then, thousands of new compounds have been synthesized with non-carbon atoms incorporated in fullerenes, sometimes actually caged inside them. Nanotubes exhibit promising characteristics for various applications. They are excellent conductors of heat and electricity, exhibit novel electrical properties, possess extreme tensile strength, and can penetrate membranes such as cell walls. Their applications include electronics, structural materials, and medicine beckon [47–49].

In 2000, the United States set the pace for nanotechnology innovation worldwide with the advent of the National Nanotechnology Initiative (NNI). The NNI today consists of individual and cooperative nanotechnology-related activities of federal agencies with a range of research and regulatory roles and responsibilities. They are working together toward the shared vision of a future in which the ability to understand and control matter at the nanoscale leads to a revolution in technology and industry that benefits society [50]. In 2006, IBM researchers succeeded in building the first electronic integrated circuit with a single carbon nanotube, heralding further advances in molecular electronics [51]. Among the applications currently available to consumers, carbon-nanotube-based LED displays are clear examples of the development in this field [52].

Nanotechnology will replace the entire manufacturing base with a new, radically more precise, radically less expensive, and radically more flexible way of making products. The aim is not simply to replace today's computer chip-making plants but also to replace the assembly lines for cars, televisions, telephones, books, surgical tools, missiles, bookcases, airplanes, tractors, and others. The objective is a pervasive change in manufacturing, a change that will leave virtually no product untouched.

Economic progress and military readiness in the 21st century will depend fundamentally on maintaining a competitive position in nanotechnology [8, 11, 14, 15]. The demand for low-cost, flexible, lightweight, robust, and low-power electronic devices that are fabricated on flexible materials using commercial and low-cost techniques, suitable for mass production, in addition to ensuring high performance, has become a challenge [53].

Therefore, several researchers have tried to use inexpensive, reliable, durable, and environmental substrates for the implementation of numerous electronic devices and RF structures, such as RFID tags, and various antenna designs used in sensor nodes for several applications [54]. Most devices used in daily life contain at least one or more antennas, such as cell phones, routers, radio communication devices, and tracking items. Antennas are also used in biomedical devices for sensing and imaging [55–60].

New antenna design technologies are being developed to allow an antenna to rapidly change its pattern in response to changes in the direction of arrival of received signal and frequency, considering the size, fabrication, impedance bandwidth, radiation pattern, and ability to integrate into different applications for numerous devices [31, 33]. There are numerous types of antenna models, such as wire (dipole, loop), traveling wave (Yagi, helical), reflector (parabolic), log-periodic (bow-tie), aperture (horn), and microstrip antenna. Each configuration of these antenna models is unique and used for certain applications. Therefore, researchers try to invent new antenna designs by combining common types of known antenna configurations to obtain new designs with high performance and overcome the disadvantages of each type [61].

Currently, antenna implementation is being developed to provide many advantages such as lightweight, low volume, low cost, low profile, and small size on the condition of the ease of the fabrication process. In addition, the performance metrics of antennas are required to be enhanced including higher gain and flexible design. Among several types of antenna configurations, microstrip antennas are considered most suitable for various practical applications, as they have several advantages.

Compared with conventional antennas, microstrip antennas have more advantages and better prospects, such as providing dual and circular polarizations, dual-frequency operation, feedline flexibility, and beam scanning omnidirectional patterning [62–64]. The major interest today is developing small-sized multiband and multifunction antennas that could serve different applications. Such antennas should be discreet and flexible to enable them to be used in the mass consumer market, using commercial printing techniques such as screen and inkjet. In addition, the strong growth in demand for wireless devices suggests that the number of antennas in use may be increased to one antenna per object in the world. For example, RFID will be integrated into each product, container, pet, banana, toy, etc. [58, 60].

1.3 Microstrip Antennas

Microstrip antennas are frequently used in today's wireless communication systems, due to their low profile and ability to mount to the walls of buildings, to the fuselages of airplanes, or to the reverse sides of mobile phones. Moreover, microstrip antennas are fabricated relatively simply and easily. Also, microstrip antennas can be simply integrated directly into microwave circuits [62]. Therefore, they are extremely compatible with embedded antennas in handheld wireless devices such as cellular phones, pagers, etc. The telemetry and communication antennas on missiles need to be thin and conformal and they are often microstrip patch antennas [65].

The basic design form of a microstrip antenna can take any possible shape such as square, rectangular, circular, triangular, elliptical, or any other common shape, as shown in Fig. 1.2. It consists of a radiating patch on one side of a dielectric substrate and a ground plane on the other side. The patch is generally made of conducting materials such as copper, gold, or silver [62, 65]. The following sections present the basic types of microstrip antennas and common feeding techniques, followed by basic calculations and examples of microstrip antennas in different applications.

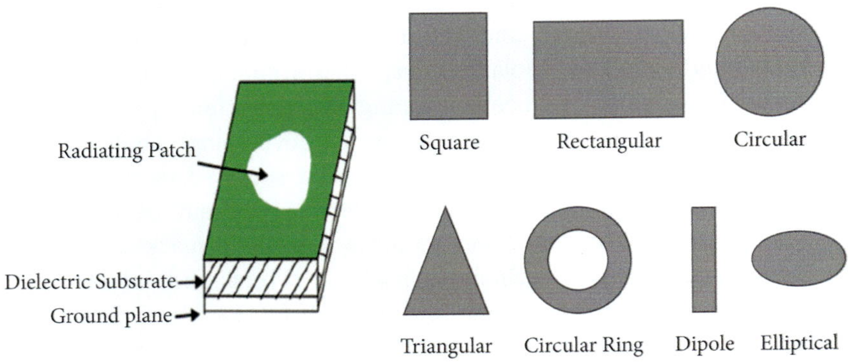

Figure 1.2 Basic structure of microstrip antennas with typical shapes [62, 65].

1.4 Microstrip Antenna Categories and Applications

Most of the research on microstrip antennas focused on methods to enhance their performance, so they are divided into categories based on the requirement and number of physical parameters. They have many geometrical shapes and dimensions to integrate easily into different applications [62, 64, 65]. Microstrip antennas are divided into main four basic categories, namely, patch, dipole, slot, and traveling wave. Such variations of microstrip antenna categories nominate them to be a part of lots of applications in numerous fields, such as medical applications, satellites, military (rockets and aircraft missiles), and others.

Wide use of microstrip antennas instead of conventional antennas is expected, due to easy design, analysis, and the ability of microstrip antennas to meet all requirements [31, 33, 55, 56, 66]. Some applications of microstrip antennas that are useful include mobile communication systems, which require small-size, low-cost, low-profile antennas.

The IEEE 802.16 standard is called Worldwide Interoperability for Microwave Access (WiMAX). It can (theoretically) reach up to a 30-mile radius with a data rate of 70 Mbps. WiMAX operates at three resonant modes (2.7, 3.3, and 5.3 GHz). Therefore, a microstrip multiband antenna is used to compliant WiMAX communication equipment [67]. Another system such as the global positioning system (GPS) requires microstrip patch antennas that have a substrate with high permittivity material, low cost, and lightweight [68].

Furthermore, vehicles depending on GPS mostly use a circularly polarized antenna with compact size. It is expected that many GPS receivers will be used by the general population for land vehicles, aircraft, and maritime vessels to find their position accurately [69]. Radar systems can also be used for detecting moving targets such as people and vehicles. These require a low-profile, lightweight antenna [70, 71]. Moreover, satellite communication systems require circularly polarized radiation pattern antennas, which can be realized easily using either square or circular patches with one or two feed points [26, 32, 72, 73].

Another kind of application requires antenna designs with very high directive characteristics to meet the demands of short and long-distance links by increasing the electrical size of the antenna, such as radio frequency identification and rectenna [55, 74]. RFIDs are involved in different applications. They take part in mobile communication, logistics, manufacturing, transportation, and healthcare. RFID systems generally use frequencies between 30 Hz and 5.8 GHz depending on their applications. Basically, an RFID system has a tag or transponder and a transceiver or reader [33, 55, 66]. Rectenna, or rectifying antenna, is a special type of antenna that is used to convert microwave energy into DC power [74, 75].

In telemedicine applications, an antenna operates at 2.45 GHz. A wearable microstrip antenna is suitable for wireless body area networks (WBAN). The required antenna should have a high gain and front-to-back ratio compared to other antennas.

Moreover, in some medical applications, such as the treatment of malignant tumors, it has been found that microwave energy is the most effective way of inducing hyperthermia. The radiator to be used for this purpose should be lightweight, compact, and easy to handle.

There is also a simple operation that goes on with medical instruments, such as coupled microstrip lines that are separated with a flexible separation are used to measure the temperature inside the human body [22].

References

1. K. S. Rao, P. V. Nikitin, and S. F. Lam, "Antenna design for UHF RFID tags: A review and a practical application," *IEEE Transactions on Antennas and Propagation*, vol. 53, no. 12, pp. 3870–3876, 2005.
2. L. Yang, A. Rida, R. Vyas, and M. M. Tentzeris, "RFID tag and RF structures on a paper substrate using inkjet-printing technology," *IEEE Transactions on Microwave Theory and Techniques*, vol. 55, no. 12, pp. 2894–2901, 2007.
3. T. Zhang, "Methods for fabricating printed electronics with high conductivity and high resolution," Ph.D. dissertation, The University of Western Ontario, 1842.
4. M. Swan, "Sensor mania! the internet of things, wearable computing, objective metrics, and the quantified self 2.0," *Journal of Sensor and Actuator Networks*, vol. 1, no. 3, pp. 217–253, 2012.
5. Y. Chen, J. Au, P. Kazlas, A. Ritenour, H. Gates, and M. McCreary, "Electronic paper: Flexible active-matrix electronic ink display," *Nature*, vol. 423, no. 6936, pp. 136–136, 2003.
6. U. Bakshi and A. Bakshi, "*Antenna and Wave Propagation*," Technical Publications, 2009.
7. R. G. S. Goh, "Carbon nanotubes for organic electronics," Ph.D. thesis, Queensland University of Technology, 2008.
8. R. J. Hamers, "Flexible electronic futures," *Nature*, vol. 412, no. 6846, pp. 489–490, 2001.
9. J. J. Miniet, "Double-sided flexible electronic circuit module," Jan. 28, 1986, US Patent 4,567,543.
10. D. R. Cairns and G. P. Crawford, "Electromechanical properties of transparent conducting substrates for flexible electronic displays," *Proceedings of the IEEE*, vol. 93, no. 8, pp. 1451–1458, 2005.

11. B. Lahey, A. Girouard, W. Burleson, and R. Vertegaal, "Paperphone: Understanding the use of bend gestures in mobile devices with flexible electronic paper displays," in *Proceedings of the SIGCHI Conference on Human Factors in Computing Systems.* ACM, 2011, pp. 1303–1312.
12. S. Park and S. Jayaraman, "Smart textiles: Wearable electronic systems," *MRS Bulletin*, vol. 28, no. 08, pp. 585–591, 2003.
13. C. Pang, C. Lee, and K.-Y. Suh, "Recent advances in flexible sensors for wearable and implantable devices," *Journal of Applied Polymer Science*, vol. 130, no. 3, pp. 1429–1441, 2013.
14. W. S. Wong, M. L. Chabinyc, T.-N. Ng, and A. Salleo, "Materials and novel patterning methods for flexible electronics," in *Flexible Electronics.* Springer, 2009, pp. 143–181.
15. J. Bae, M. K. Song, Y. J. Park, J. M. Kim, M. Liu, and Z. L. Wang, "Fiber supercapacitors made of nanowire-fiber hybrid structures for wearable/flexible energy storage," *Angewandte Chemie International Edition*, vol. 50, no. 7, pp. 1683–1687, 2011.
16. C. M. Janik, "Flexible wearable computer," Aug. 22 2000, US Patent 6,108,197.
17. S. Hong and S. Myung, "A flexible approach to mobility," *Nature Nanotechnology*, vol. 2, pp. 207–208, 2007.
18. B. Kavoossi and E. Otaniemi, "Active imaging and visual information on paper and paper-based products," Printed Microsystems on Paper, pp. 1–106, 2001.
19. T. T. Bakhishev, "Solution-processed inorganic electronics," Ph.D. dissertation, University of California, Berkeley, 2011.
20. R. P. Feynman, "There's plenty of room at the bottom [data storage]," *Journal of Microelectromechanical Systems*, vol. 1, no. 1, pp. 60–66, 1992.
21. J. Kim, Z. Wang, and W. S. Kim, "Stretchable RFID for wireless strain sensing with silver nano ink," *IEEE Sensors Journal*, vol. 14, no. 12, pp. 4395–4401, 2014.
22. G. Cohn, D. Morris, S. Patel, and D. Tan, "Humantenna: Using the body as an antenna for real-time whole-body interaction," in *Proceedings of the SIGCHI Conference on Human Factors in Computing Systems.* ACM, 2012, pp. 1901–1910.
23. T. Rai, P. Dantes, B. Bahreyni, and W. S. Kim, "A stretchable RF antenna with silver nanowires," *IEEE Electron Device Letters*, vol. 34, no. 4, pp. 544–546, 2013.

24. L. Song, A. C. Myers, J. J. Adams, and Y. Zhu, "Stretchable and reversibly deformable radio frequency antennas based on silver nanowires," *ACS Applied Materials & Interfaces*, vol. 6, no. 6, pp. 4248–4253, 2014.

25. H. Singh, R. Chandini, and R. M. Jha, "Low profile conformal antenna arrays on high impedance substrate," in *Low Profile Conformal Antenna Arrays on High Impedance Substrate*. Springer, 2016, pp. 1–50.

26. D. Betancourt and J. Castan, "Printed antenna on flexible low-cost pet substrate for uhf applications," *Progress in Electromagnetics Research C*, vol. 38, pp. 129–140, 2013.

27. H. R. Khaleel, H. M. Al-Rizzo, and D. G. Rucker, "Compact polyimide-based antennas for flexible displays," *Journal of Display Technology*, vol. 8, no. 2, pp. 91–97, 2012.

28. A. C. Durgun, C. A. Balanis, C. R. Birtcher, and D. R. Allee, "Design, simulation, fabrication and testing of flexible bow-tie antennas," *IEEE Transactions on Antennas and Propagation*, vol. 59, no. 12, pp. 4425–4435, 2011.

29. L. Marnat and A. Shamim, "Liquid crystal polymer (LCP) based antenna for flexible system on package (SOP) applications," in *Antenna Technology and Applied Electromagnetics (ANTEM), 2012 15th International Symposium on*. IEEE, 2012, pp. 1–4.

30. A. Rida, L. Yang, R. Vyas, S. Bhattacharya, and M. M. Tentzeris, "Design and integration of inkjet-printed paper-based UHF components for RFID and ubiquitous sensing applications," in *Microwave Conference, 2007*. European. IEEE, 2007, pp. 724–727.

31. K. Ray and Y. Ranga, "CPW-fed modified rectangular printed monopole antenna with slot," in *2007 SBMO/IEEE MTT-S International Microwave and Optoelectronics Conference*, Salvador, Brazil, 2007, pp. 79–81.

32. R. Waterhouse, *Printed Antennas for Wireless Communications*. Wiley Online Library, 2007.

33. J. Sidén and H.-E. Nilsson, *RFID Antennas-Possibilities and Limitations*. INTECH Open Access Publisher, 2010.

34. A. Mansour, N. Shehata, B. Hamza, and M. Rizk, "Efficient design of flexible and low cost paper-based inkjet-printed antenna," *International Journal of Antennas and Propagation*, Article ID 845042, 2015.

35. K. E. Drexler, "Nanotechnology: From Feynman to funding," *Bulletin of Science, Technology & Society*, vol. 24, no. 1, pp. 21–27, 2004.

36. N. Taniguchi et al., "On the basic concept of nanotechnology," in *Proc. Intl. Conf. Prod. Eng. Tokyo, Part II, Japan Society of Precision Engineering*, 1974, pp. 18–23.
37. J. V. Barth, G. Costantini, and K. Kern, "Engineering atomic and molecular nanostructures at surfaces," *Nature*, vol. 437, no. 7059, pp. 671–679, 2005.
38. S. Sahoo, S. Parveen, and J. Panda, "The present and future of nanotechnology in human healthcare," *Nanomedicine: Nanotechnology, Biology and Medicine*, vol. 3, no. 1, pp. 20–31, 2007.
39. F. A. Buot, "Mesoscopic physics and nanoelectronics: nanoscience and nanotechnology," *Physics Reports*, vol. 234, no. 2, pp. 73–174, 1993.
40. D. Goldhaber-Gordon, M. S. Montemerlo, J. C. Love, G. J. Opiteck, and J. C. Ellenbogen, "Overview of nanoelectronic devices," *Proceedings of the IEEE*, vol. 85, no. 4, pp. 521–540, 1997.
41. C. Bai, *Scanning Tunneling Microscopy and Its Application*. Springer Science & Business Media, 2000, vol. 32.
42. V. Balzani, A. Credi, F. M. Raymo, and J. F. Stoddart, "Artificial molecular machines," *Angewandte Chemie International Edition*, vol. 39, no. 19, pp. 3348–3391, 2000.
43. "Doing mm," http://wise-nano.org/w/Doing MM.
44. R. C. Merkle, "Are we prepared for the nanotechnology revolution?" *Proceedings of the IEEE*, vol. 88, no. 1, pp. 107–108, 2000.
45. "Buckyballs," http://www.chemheritage.org/discover/online-resources/chemistry-in-history/themes/microelectronics-and-nanotechnology/ smalley-curl-kroto.aspx.
46. S. Iijima et al., "Helical microtubules of graphitic carbon," *Nature*, vol. 354, no. 6348, pp. 56–58, 1991.
47. B. I. Yakobson and R. E. Smalley, "Fullerene nanotubes: C 1,000,000 and beyond: Some unusual new molecules—long, hollow fibers with tantalizing electronic and mechanical properties—have joined diamonds and graphite in the carbon family," *American Scientist*, vol. 85, no. 4, pp. 324–337, 1997.
48. P. G. Collins and P. Avouris, "Nanotubes for electronics," *Scientific American*, vol. 283, no. 6, pp. 62–69, 2000.
49. W. Liang, X. J. Wang, S. Yokojima, and G. Chen, "Electronic structures and optical properties of open and capped carbon nanotubes," *Journal of the American Chemical Society*, vol. 122, no. 45, pp. 11 129–11 137, 2000.

50. E. Fisher and R. L. Mahajan, "Contradictory intent? Us federal legislation on integrating societal concerns into nanotechnology research and development," *Science and Public Policy*, vol. 33, no. 1, pp. 5–16, 2006.

51. Z. Chen, J. Appenzeller, Y.-M. Lin, J. Sippel-Oakley, A. G. Rinzler, J. Tang, S. J. Wind, P. M. Solomon, and P. Avouris, "An integrated logic circuit assembled on a single carbon nanotube," *Science*, vol. 311, no. 5768, pp. 1735–1735, 2006.

52. P. Avouris, M. Freitag, and V. Perebeinos, "Carbon-nanotube photonics and optoelectronics," *Nature Photonics*, vol. 2, no. 6, pp. 341–350, 2008.

53. L. Yang and M. M. Tentzeris, "Design and characterization of novel paper-based inkjet-printed RFID and microwave structures for telecommunication and sensing applications," in *2007 IEEE/MTT-S International Microwave Symposium*. IEEE, 2007, pp. 1633–1636.

54. T. Mishima, N. Abe, K. Tanaka, and H. Taki, "Toward construction of a mobile system with long-range RFID sensors," in *Cybernetics and Intelligent Systems, 2004 IEEE Conference on, vol. 2*. IEEE, 2004, pp. 960–965.

55. Z. Konstas, A. Rida, R. Vyas, K. Katsibas, N. Uzunoglu, and M. M. Tentzeris, "A novel "green" inkjet-printed Z-shaped monopole antenna for RFID applications," in *2009 3rd European Conference on Antennas and Propagation*, Berlin, Germany, 2009, pp. 2340–2343.

56. Y. Kawahara, S. Hodges, B. S. Cook, C. Zhang, and G. D. Abowd, "Instant inkjet circuits: Lab-based inkjet printing to support rapid prototyping of ubicomp devices," in *Proceedings of the 2013 ACM International Joint Conference on Pervasive and Ubiquitous Computing*. ACM, 2013, pp. 363–372.

57. D. Masse, "RFID handbook: Fundamentals and applications in contactless smart cards and identification second edition," *Microwave Journal*, vol. 47, no. 10, pp. 168–169, 2004.

58. C. Loebbecke and J. W. Palmer, "RFID in the fashion industry: Kaufhof department stores AG and Gerry Weber International AG, fashion manufacturer," *MIS Quarterly Executive*, vol. 5, no. 2, 2006.

59. L. Ricciardi, I. Pitz, S. F. Al-Sarawi, V. Varadan, and D. Abbott, "Investigation into the future of RFID in biomedical applications," in *Microtechnologies for the New Millennium 2003. International Society for Optics and Photonics, 2003*, pp. 199–209.

60. X. Zhu, S. K. Mukhopadhyay, and H. Kurata, "A review of RFID technology and its managerial applications in different industries,"

Journal of Engineering and Technology Management, vol. 29, no. 1, pp. 152–167, 2012.

61. http://www.antenna-theory.com/antennas/main.php.
62. I. Singh and D. V. S. Tripathi, "ISSN: 2229-6093 micro strip patch antenna and its applications: A survey," 2011.
63. R. Garg, *Microstrip Antenna Design Handbook*. Artech House, 2001.
64. H. Koa, Y. J. Kimb, M. Q. Fengb, and F. De Flaviisa, "Pin switched microstrip slot antenna array for 3D microwave reflection tomography," EEECS Department, University of California.
65. P. Ramachandran, T. Keshav, L. Minz, V. Parupalli, and S. Chakravarty, "Antenna design, simulation and fabrication," Dept of Electronics and Computer Science Engineering, Visvesvaraya National Institute of Technology, Nagpur, 2007.
66. K. Finkenzeller, *RFID Handbook: Fundamentals and Applications in Contactless Smart Cards and Identification*, Chichester, 2003.
67. W.-S. Chen and Y.-H. Yu, "Dual-band printed dipole antenna with parasitic element for WiMAX applications," *Electronics Letters*, vol. 44, no. 23, pp. 1338–1339, 2008.
68. D. M. Pozar and S. M. Duffy, "A dual-band circularly polarized aperture-coupled stacked microstrip antenna for global positioning satellite," *IEEE Transactions on Antennas and Propagation*, vol. 45, no. 11, pp. 1618–1625, 1997.
69. S. K. Patel and Y. Kosta, "E-shape microstrip patch antenna design for GPS application," in *2011 Nirma University International Conference on Engineering*. IEEE, 2011, pp. 1–4.
70. S. Stuchly, A. Thansandote, J. Mladek, and J. Townsend, "A doppler radar velocity meter for agricultural tractors," *IEEE Transactions on Vehicular Technology*, vol. 27, no. 1, pp. 24–30, 1978.
71. H. Moheb, L. Shafai, and M. Barakat, "Design of 24 GHz microstrip travelling wave antenna for radar application," in *Antennas and Propagation Society International Symposium, 1995*. AP-S. Digest, vol. 1. IEEE, 1995, pp. 350–353.
72. T.-L. Zhang, Z.-H. Yan, L. Chen, and Y. Song, "A compact dual-band CPW-fed planar monopole antenna for WLAN applications," *Journal of Electromagnetic Waves and Applications*, vol. 22, no. 14–15, pp. 2097–2104, 2008.
73. K. Zhao, S. Zhang, and S. He, "Closely-located MIMO antennas of tri- band for WLAN mobile terminal applications," *Journal of Electromagnetic Waves and Applications*, vol. 24, no. 2–3, pp. 363–371, 2010.

74. M. Al-Husseini, A. Haskou, N. Rishani, and K. Y. Kabalan, "Textile-based rectennas," *WIT Transactions on State-of-the-art in Science and Engineering*, vol. 82, p. 187, 2014.

75. V. Bhatnagar and P. Owende, "Energy harvesting for assistive and mobile applications," *Energy Science & Engineering*, vol. 3, no. 3, pp. 153–173, 2015.

Chapter 2

Printing Techniques

Ahmed Mansour and Effat Samir

Department of Electrical Engineering, Alexandria University, Alexandria, Egypt

ahmed_mansour@mena.vt.edu

Printed electronics comprise a set of printing methods used to create electrical devices on various substrates. Printing typically uses common printing equipment suitable for defining patterns on materials, such as lithography, spin coating screen, and inkjet. Printed electronics are expected to facilitate widespread, very low-cost, low-performance electronics for applications such as flexible displays, smart labels, decorative and animated posters, and interactive clothes [1–5].

Therefore, the development of printing methods and ink materials are essential fields that take the intention [6]. The most important parameters of the printing technique are cost and simplicity per volume fabrication. This chapter proposes the common printing methods that are used for electronics printing, showing their operation concepts and the main parameters that affect the fabrication process.

Flexible Inkjet-Printed Antennas: Design, Fabrication, and Applications
Edited by Nader Shehata

ISBN 978-981-4968-87-4 (Hardcover), TBA (eBook)
www.jennystanford.com

DOI: 10.1201/9781003774754-2

2.1 Lithography

Lithography is one of the oldest printing methods that is based on the mapping of an image onto the surface of a lithographic plate. Lithography is used presently to produce posters, maps, books, newspapers, etc. [7, 8]. The semiconductor fabrication is provided by the basic ideas of lithographic methods, as it is a convenient, high-speed printing technique developed within the graphical industry. It is also a conceivable production process for electronics on flexible substrates [4, 9].

Figure 2.1 shows a reel-to-reel (R2R) high-speed printing method (or o set printing method). R2R is just like classical lithography, which is based on the repulsion between hydrophobic and hydrophilic substances [10].

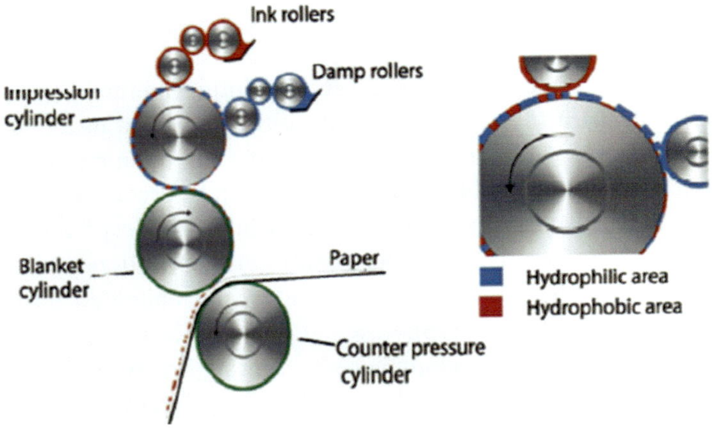

Figure 2.1 Illustration of the R2R printing technique [10].

An impression cylinder carries the image to be printed, defined from hydrophobic and hydrophilic areas. Ink and water are set from this cylinder to an intermediate blanket cylinder, which, in turn, transfers the image to the paper [10].

2.2 Spin Coating

Spin coating generally involves applications that require thin film (in the range of a few nanometers to a few micrometers) evenly across the surface of a substrate by coating a solution of the desired material in a solvent (ink) while it is rotating. Figure 2.2 shows the basic setup of spin coating. This technique is used in inorganic semiconductor printing, where it is mostly employed to deposit organic photoresist on silicon wafers. The resulting film characteristics depend on different parameters that affect the spin-coating process. Such parameters are time, speed, plate temperature, solvent evaporation rate, and properties of the fluid to be spun [4, 5, 11]. The normal range of spin coating rpm spans from 600 to 6000 rpm, and it is an important parameter that affects the thicknesses [11, 13].

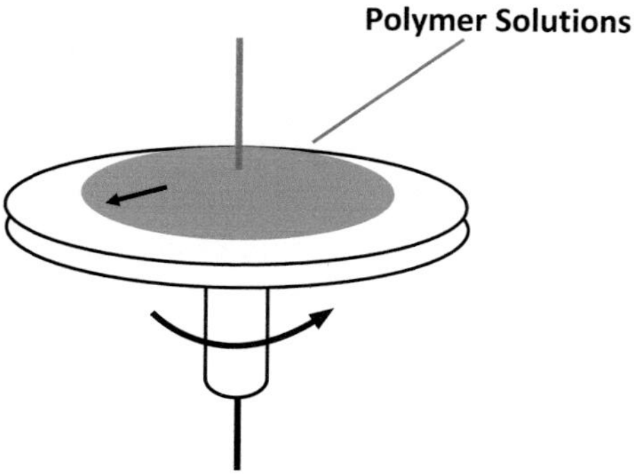

Figure 2.2 Illustration of spin coating [12].

The maximum thickness also depends on the concentration of the material that can be dissolved in the solvent. For high solubility materials (100 mg/ml or higher), thicknesses of >1 μm can be achieved. Meanwhile, for some low solubility conjugated polymers (a few mg/ml), the maximum thickness might be limited to 20 nm or so. At low concentrations, the thickness of a film is approximately linearly dependent upon the concentration of the material in the ink [4, 13].

For most standard spin coating techniques, the objective is to keep the substrate spinning until the film is fully dry [14]. As such, this will mainly depend upon the boiling point and vapor pressure of the solvent that is used and also on the ambient conditions (temperature and humidity) that the spin coating is performed in. For higher boiling point (and/or lower vapor pressure) solvents, drying may take considerably longer time (up to ten minutes in some cases), and therefore, these solvents are most commonly used either as additives or with additional drying steps (covered in later sections). The large material waste is often mentioned as a disadvantage of this deposition technique [12].

2.3 Reactive/Non-reactive Patterning

Patterning techniques that require specific chemical polymers to adapt the material to be deposited and patterned are called reactive techniques. For example, photo patterning or photo-chemical patterning of conjugated polymers relies on the chemical changes that some polymers undergo when they are exposed to light [15–18].

Moreover, when patterning methods do not involve complex chemical reactions, they are called non-reactive techniques. This technique is commonly used for different applications, and the most common examples are screen printing, soft lithography (called soft because it uses elastomeric materials such as polydimethylsiloxane "PDMS"), and inkjet printing [10, 19, 20].

2.4 Inkjet Printing

Inkjet printing is a very promising technology in the scientific research field because of its main characteristics such as high accuracy, non-contact patterning, simplicity, high speed, and high-quality deposition on both small and large areas [10]. Also, the inkjet technique has a low cost and is suitable for printing on flexible substrates. In addition, this method is convenient for the deposition of inks, such as polymer solution and pure solvents or solvent mixtures, onto various surfaces and substrates. Some examples of functional materials include metal inks, conductive and semiconductive polymers, surface coating, proteins, and nanoparticles [21].

The basic idea of inkjet printing is to print graphical images by ringing tiny droplets of ink onto a substrate. There are several ways to create the droplets and to control where the droplets should land on the substrate. The inkjet technology can be divided into two main classes: continuous inkjet and drop-on-demand inkjet (DOD), as shown in Fig. 2.3.

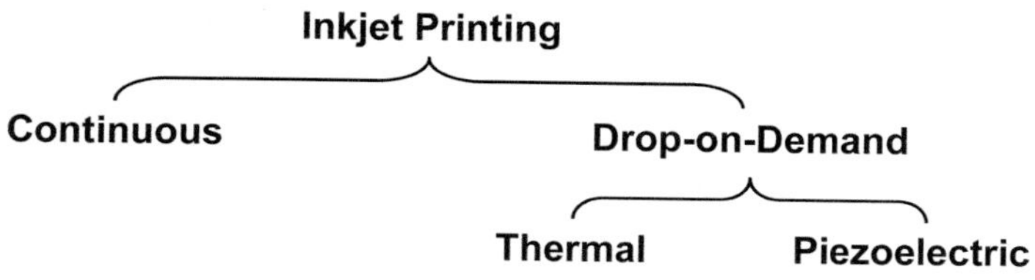

Figure 2.3 Diagram of the available inkjet printing technologies.

2.4.1 Continuous Inkjet Printing Techniques

Continuous inkjet printing techniques are mainly used in high-volume applications, such as industrial coding, marking, and labeling. This operation requires a continuous generation of ink stream from a pressurized fluid reservoir [10].

2.4.2 Drop-on-Demand Inkjet Printing Techniques

DOD inkjet printing technology is used in most printers. Because of the smaller drop size and the higher positioning accuracy of DOD inkjet printers, they are the key to the direct printing of the inks. This approach eliminates the complexity of drop charging and defection hardware. Most drop-on-demand solutions fall into one of the two main technology categories, i.e., thermal inkjet and piezoelectric inkjet. These technologies are used in both office/home printers and more advanced printers for professional and industrial uses [10].

2.4.2.1 Piezoelectric inkjet

This inkjet technology is based on piezoelectric crystals confined in each printer nozzle. A voltage pulse applied to the piezoelectric crystal causes it to change shape and size. This creates a pressure pulse in the liquid ink and a drop of ink is released through the small orifice of the nozzle [10]. The mounting of the crystal within the nozzle can be done in several ways. In some printers' heads, the crystal constitutes one or several sides of the walls of the nozzle, as shown in Fig. 2.4. As the crystal is made to pulse, the nozzle channel width varies. The crystal is mounted at the top end of the nozzle rather than modifying the length of the ink column in the nozzle. Companies such as XAAR, Brother, Micro Fab Tech, and Seiko Epson have their versions of this technology [10].

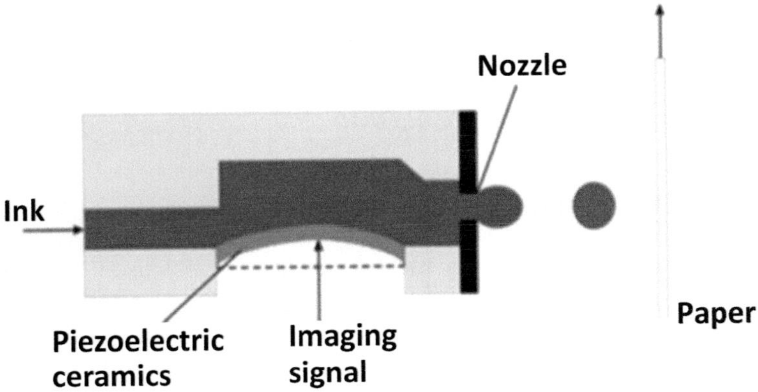

Figure 2.4 Illustration of the principle of a piezoelectric inkjet printing head [10].

2.4.2.2 Thermal inkjet

In a thermal inkjet print head, the imaging signals are connected to small heater elements located within the nozzles. To make a droplet out of a nozzle, its heater is activated. The rapid local heating results in the creation of a vapor bubble inside the nozzle. As the bubble expands, the pressure within the nozzle will increase and ink will be forced out through the orifice, forming a drop. In this way, drops are "jetted" out of the nozzle toward the substrate. What separates different types of thermal inkjet systems is mainly the position of the heater element. It can be located directly above the orifice, or in the walls of the nozzle, as shown in Fig. 2.5. Thermal inkjet technology dominates the home and office markets, and the development is driven by big players in these markets such as Canon and Hewlett Packard [10, 22].

Thermal inkjet is considered the most common inkjet technique and it dominates the market for home users. In this technique, the ink drops are ejected from a nozzle by the growth and collapse of a vapor bubble on the top surface of a small heater located near the nozzle. With a current pulse of less than a few microseconds through the heater, the ink's temperature rises to the critical value for bubble nucleation. When the nucleation occurs, a vapor bubble expands rapidly to force the ink to form a drop outside the nozzle. The whole process of bubble formation and collapse takes place in less than 10 seconds. The ink then refills back into the chamber and the process is ready to restart [23].

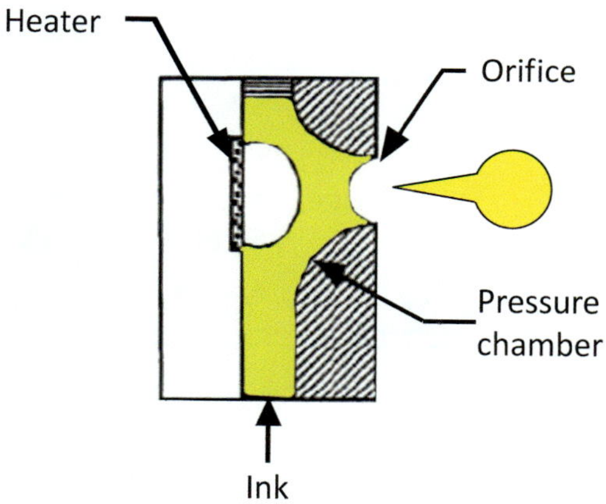

Figure 2.5 Schematic illustration of a thermal inkjet [21].

2.5 Screen Printing Procedures

Screen printing is a kind of stencil printing mechanism, which is also known as silk screen printing. Screen printing is used to produce massive production of graphics, such as posters or writing on a printed circuit board. This technique is simple and compatible with various substrate materials and surfaces [10]. The screen-printing process basically depends on the exposed substrate to UV light. The emulsion is used to cover the parts of the screen, which are not supposed to let any paint through. The uncovered areas of the screen can let paint pass through it and thus be transferred onto the substrate. The unexposed emulsion can be washed off the screen, while the hardened emulsion will stay on, forming the stencil. An alternative method to create the stencil is to directly print the emulsion onto the screen using for example inkjet technique.

Screen printing procedures start with placing the screen, and frame above the substrate, either directly in contact with the substrate or a few millimeters above it. The paint ink will be transferred from the top side of the screen to the substrate by pressing it through the screen with a squeegee, which is simply a rubber blade. The paint cannot pass through the areas covered by emulsion and the required image will be printed onto the substrate, as shown in Fig. 2.6. Screen printing depends on many parameters that control the printing process, such as thickness, squeegee angle, and resolution.

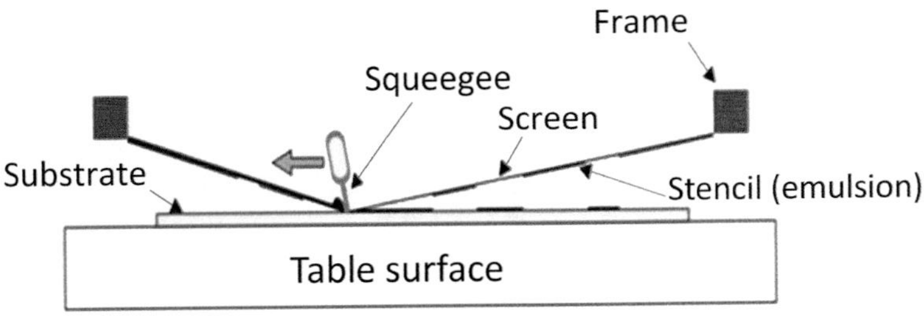

Figure 2.6 Schematic illustration of screen printing [10].

The thickness depends on the viscosity of the paint, while the emulsion, the squeegee profile, and the angle of the squeegee are in relation to the screen when swept across it. The obtainable resolution of screen printing is just like the inkjet technique, and its resolution depends on the edges of the emulsion layer [10].

Screen printing and inkjet printing are the most common printing techniques, while screen printing is the best option for designs that require mass production and a high level of vibrancy. Moreover, the ink that is used in screen printing is applied thicker than inkjet printing. Inkjet printing is best used for items that are required high amounts of detail, and orders of a smaller quantity.

References

1. J. Rogers, H. Katz et al., "Printable organic and polymeric semiconducting materials and devices," *Journal of Materials Chemistry*, vol. 9, no. 9, pp. 1895–1904, 1999.
2. M. Singh, H. M. Haverinen, P. Dhagat, and G. E. Jabbour, "Inkjet printing—process and its applications," *Advanced Materials*, vol. 22, no. 6, pp. 673–685, 2010.
3. S. H. Ko, H. Pan, C. P. Grigoropoulos, C. K. Luscombe, J. M. Fréchet, and D. Poulikakos, "All-inkjet-printed flexible electronics fabrication on a polymer substrate by low-temperature high-resolution selective laser sintering of metal nanoparticles," *Nanotechnology*, vol. 18, no. 34, p. 345202, 2007.
4. S. H. Ko, H. Pan, C. P. Grigoropoulos, J. M. Fréchet, C. K. Luscombe, and D. Poulikakos, "Lithography-free high-resolution organic transistor arrays on polymer substrate by low energy selective laser ablation of inkjet-printed nanoparticle film," *Applied Physics A*, vol. 92, no. 3, pp. 579–587, 2008.
5. K. J. Lee, B. H. Jun, T. H. Kim, and J. Joung, "Direct synthesis and inkjetting of silver nanocrystals toward printed electronics," *Nanotechnology*, vol. 17, no. 9, p. 2424, 2006.
6. H.-K. Roth, S. Sensfuss, M. Schrodner, R.-I. Stohn, W. Clemens, and A. Bernds, "Organische funktionsschichten in polymerelektronik und polymersolarzellen," *Materialwissenschaft und Werkstofftechnik*, vol. 32, no. 10, pp. 789–794, 2001.
7. R. Bassemir and A. Bean, "Inks," *Kirk-Othmer Encyclopedia of Chemical Technology*, 2004.
8. M. H. Bruno, "The printer's role in a digital printing world: A summary of digital printing systems." *Journal of Printing Science and Technology*, vol. 31, no. 5, pp. 322–339, 1994.
9. W. S. Wong, M. L. Chabinyc, T.-N. Ng, and A. Salleo, "Materials and novel patterning methods for flexible electronics," in *Flexible Electronics*. Springer, 2009, pp. 143–181.
10. R. Mannerbro and M. Ranlöf, "Inkjet and screen printed electrochemical organic electronics," 2007.
11. Y.-Y. Huang and K.-S. Chou, "Studies on the spin coating process of silica films," *Ceramics International*, vol. 29, no. 5, pp. 485–493, 2003.

12. S.-S. Lee, J.-D. Hong, C. H. Kim, K. Kim, J. P. Koo, and K.-B. Lee, "Layer-by-layer deposited multilayer assemblies of ionene-type polyelectrolytes based on the spin-coating method," *Macromolecules*, vol. 34, no. 16, pp. 5358–5360, 2001.
13. L. L. Spangler, J. M. Torkelson, and J. S. Royal, "Influence of solvent and molecular weight on thickness and surface topography of spin-coated polymer films," *Polymer Engineering & Science*, vol. 30, no. 11, pp. 644–653, 1990.
14. Y. Garsany, I. L. Singer, and K. E. Swider-Lyons, "Impact of film drying procedures on RDE characterization of Pt/VC electrocatalysts," *Journal of Electroanalytical Chemistry*, vol. 662, no. 2, pp. 396–406, 2011.
15. M. Han, M. S. Rahman, J.-S. Lee, D. Khim, D.-Y. Kim, and J.-W. Park, "Surface-grafted rodlike polymers: Adaptive self-assembled monolayers and rapid photo-patterning of surfaces," *Chemistry of Materials*, vol. 23, no. 15, pp. 3517–3524, 2011.
16. M. J. Lochhead and P. Yager, "Injection molding with microstructure from reservoirs," March 21, 2000, US Patent 6,039,897.
17. M. L. Renak, G. C. Bazan, and D. Roitman, "Microlithographic process for patterning conjugated emissive polymers," *Advanced Materials*, vol. 9, no. 5, pp. 392–395, 1997.
18. Y.-T. Tao, K. Pandian, and W.-C. Lee, "Microfabrication of interdigitated polyaniline/polymethylene patterns on a gold surface," *Langmuir*, vol. 14, no. 21, pp. 6158–6166, 1998.
19. G. Kumar and K. N. Prabhu, "Review of non-reactive and reactive wetting of liquids on surfaces," *Advances in Colloid and Interface Science*, vol. 133, no. 2, pp. 61–89, 2007.
20. T. Granlund, T. Nyberg, L. Stolz Roman, M. Svensson, and O. Inganäs, "Patterning of polymer light-emitting diodes with soft lithography," *Advanced Materials*, vol. 12, no. 4, pp. 269–273, 2000.
21. I. A. Grimaldi, "Realization and optimization of organic electronic devices manufactured by innovative technique: Inkjet printing," 2011.
22. B. Ballarin, A. Fraleoni-Morgera, D. Frascaro, S. Marazzita, C. Piana, and L. Setti, "Thermal inkjet microdeposition of PEDOT:PSS on ITO-coated glass and characterization of the obtained film," *Synthetic Metals*, vol. 146, no. 2, pp. 201–205, 2004.
23. H. P. Le, "Progress and trends in ink-jet printing technology," *Journal of Imaging Science and Technology*, vol. 42, no. 1, pp. 49–62, 1998.

Chapter 3

Flexible Inkjet-Printed Antenna Fabrication, Design, and Measurement Procedures

Ahmed Mansour,[a] Mohamed Rizk,[a] and Nader Shehata[b]

[a]*Department of Electrical Engineering, Alexandria University, Alexandria, Egypt*
[b]*Department of Engineering Mathematics and Physics, Alexandria University, Alexandria, Egypt*

ahmed_mansour@mena.vt.edu, mrizk@mena.vt.edu, nader.shehata@alexu.edu.eg

Advanced technology trends underscore the emergence and increasing importance of flexible, portable, and low-energy consuming devices, such as wearable and autonomous devices, which led to talk about cost as a significant parameter [1]. The trend of fabricating electronic circuits and RF structures on flexible substrates is based on a suitable selection of different parameters. This chapter proposes experimental setup and main parameters that affect printing flexible designs such as printing technique, printing ink, and substrates that can be used. Moreover, it presents a trail to find an optimum procedure condition for printing on different substrates using the main and commercial

Flexible Inkjet-Printed Antennas: Design, Fabrication, and Applications
Edited by Nader Shehata

ISBN 978-981-4968-87-4 (Hardcover), TBA (eBook)
www.jennystanford.com

DOI: 10.1201/9781003774754-3

equipment, on condition of achieving a high performance and easy use.

3.1 Printing Technique and Printer Selection

Inkjet printing is considered the most suitable and efficient printing technique for electronic circuit and RF structure fabrication. It is more efficient than photolithography for small manufacturing scales. Inkjet printing also takes a place for this experimental work, as it is one of the most suitable printing techniques especially in research fields. It provides a fast, low-cost, simple printing technique with no requirement for a clean room. Also, inkjet printing on flexible substrates has an advantage over the traditional etched techniques, as it jets the single ink droplet from the printer's nozzles to the desired position, so there is no waste unlike the etching technique [2].

The most famous inkjet printer used in flexible electronics printing at several research labs is the Fujifilm Dimatix printer, as shown in Fig. 3.1. Such a printer has high and accurate performance. However, its price is so expensive, in addition to its quite complicated printing procedure, that it may not be used at other several labs. Therefore, a commercial, cheaper with acceptable accuracy inkjet printer is used in our work, as shown in Fig. 3.2. Brother Inkjet printer model (MFC-220J) is selected for the entire work in the thesis, as it is a common desktop printer, fast, easy to use, and does not require any special software. This printer helps in the investigations required to verify the design quickly without wasting materials at a relatively low cost compared to other material printers [3, 4].

Brother printer is based on DOD piezo inkjet technology, with CMYK (Cyan, Magenta, Yellow, and Black) cartridge. This printer costs around 100 USD, which is much cheaper compared to the famous Fujifilm Dimatix printer that has been used recently in research papers related to the printed electronics field. These printer cartridges are replaced by special external cartridges to be fitted with the used and easy filling, using a small plastic syringe. This filling is done with the help of a simple disposable filter to keep any contaminations out of the print pipeline, as shown in Fig. 3.2b. This printer provides a one-layer printed stripe, so for multiple-layer printed strips, this process is repeated by over-layered printing [4].

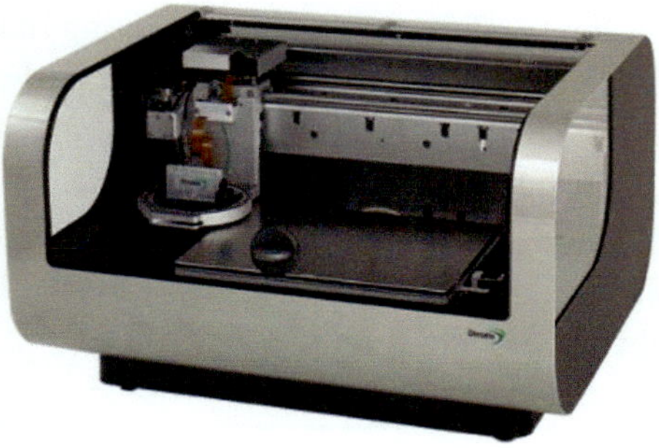

Figure 3.1 Dimatix material inkjet printer DMP-2800 series [4, 5].

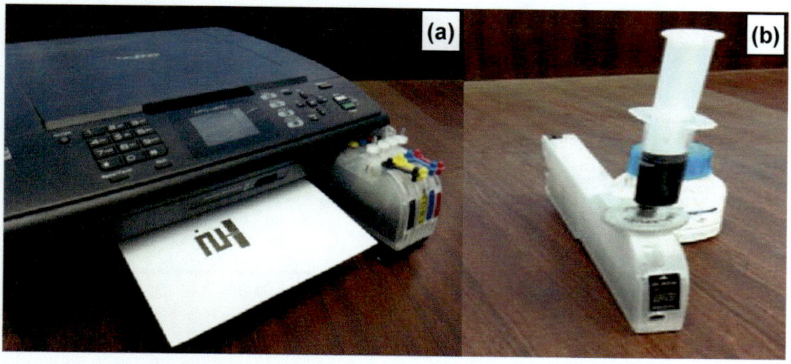

Figure 3.2 Brother inkjet printer: (a) with a printed antenna and (b) filling the cartridge with conductive silver nanoparticle ink using a syringe and filter [6].

3.2 Substrate Selection

The new trend of printing electronic circuits on flexible substrates is searching for materials that are commercially available, cheap, and environmentally friendly. However, any material can be used as a substrate through the appropriate coating process, depending on the easier fabrication process, especially for direct-write methodologies such as inkjet printing [7–12].

There are some parameters that take into consideration in the coating substrates. Such parameters are thickness and type of substrate, which play a vital role in determining the characteristics of antennas, such as impedance bandwidth, radiation efficiency, etc. The dielectric constant ε_r and substrate height (h) can be varied separately, affecting bandwidth and efficiency. It can be noticed from Fig. 3.3 that the efficiency is greater for low dielectric constant substrates, and, as the height is increased, the bandwidth also increases [13]. In this thesis, typical photo paper (glossy) and PET films are used as substrates, because they are commercially available, cheap, and environmentally friendly materials.

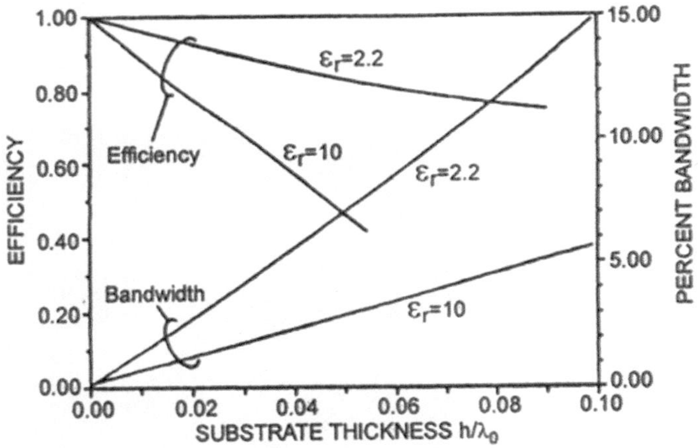

Figure 3.3 Efficiency and bandwidth versus substrate height of patch antenna [13].

Furthermore, the variations of loss tangent (tan δ) affect the input resonance resistance. It is obviously seen that as tan δ increases, the input resistance decreases. Moreover, loss tangent affects antenna operating frequency, efficiency, bandwidth, and gain. Thus, as tan δ increases, the efficiency and gain decrease, and the bandwidth increases. Therefore, lossy substrates give a wider bandwidth with reduced efficiency and lower gain, as shown in Fig. 3.4.

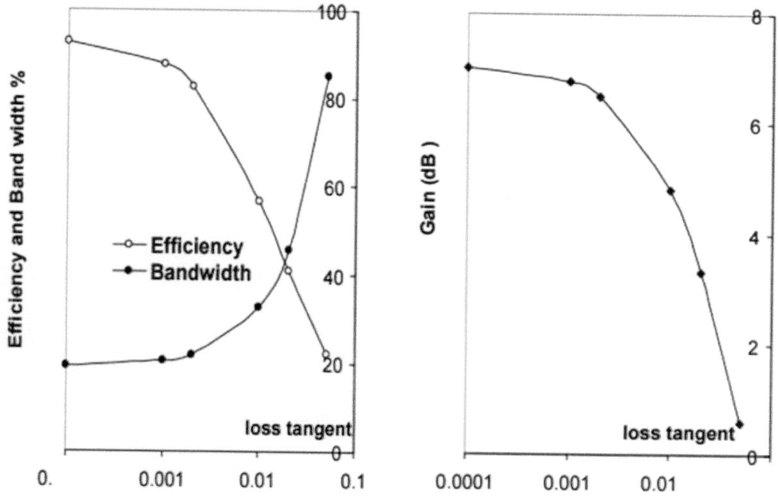

Figure 3.4 Efficiency, bandwidth, and gain versus loss tangent of patch antenna [13].

3.3 Ink Formation

A crucial element of printing technology is ink formulation. For the DOD piezoelectric inkjet technique, the selection of ink depends on the weak mechanical force generated by the piezoelectric crystal deformation on the fluid. Therefore, typical ink must have sufficiently low viscosity to allow the ejection [14]. Moreover, the optimum properties of ink that can be ejected by this printing system are summarized as follows:

- No wetting of the nozzle plate during ejection
- Compatibility with the print head and the delivery system
- Small ejecting drop volume variation among the generated droplets
- Uniform droplet velocity from droplet to droplet
- Straight droplet trajectory along the nozzle/substrate direction

This will provide good ejecting quality that meets high-performance criteria and has very narrow variability ranges

[15]. Thus, Mitsubishi silver nanoparticle ink model number NBSIJ-MU01 is nominated for experimental work in this thesis. Mitsubishi ink meets all requirements and is efficient with a Brother inkjet printer. Scanning electron microscope (SEM) shows that the silver nanoparticle ink has a mean diameter of around 100 nm, with uniform surface distribution, as shown in Fig. 3.5.

Figure 3.5 SEM image of surface morphology of the one-layer printed silver nanoparticle ink [6].

Since antennas present a significant element of all wireless devices. This chapter compares the performances of a basic microstrip antenna design, printed using the proposed low-cost printing technique. Moreover, this chapter proposes the impact of annealing and different substrates on the performance.

3.4 Impact of Temperature on Flexible Low-Cost RFID Inkjet-Printed Antenna

A simple coplanar waveguide (CPW) Z-shaped microstrip antenna design is presented with experimental and simulated analysis. This design is commonly used and it was mentioned in detail in [16] using the Fujifilm Dimatix Printer. The analysis of the surface morphology and electrical conductivity at different post-synthesis heating temperatures and multi-layer printing is considered.

This fabricated design is printed and exposed to different post-synthesis heating temperatures up to 180 °C in a furnace. The usage of silver nanoparticles shows interesting conductivity values with post-synthesis heating temperatures. It is used to enhance the performance of microstrip antenna to operate a 1.57 GHz. The proposed Z-shaped antenna has an overall area of 75 × 50 mm^2 and the full dimensions of the design are shown in Fig. 3.6. Figure 3.7 shows the fabricated Z-shaped antenna, which is connected with a sub-miniature version A(SMA) connector using silver epoxy with a ratio of 1:1 and is dried at a room temperature (25 °C).

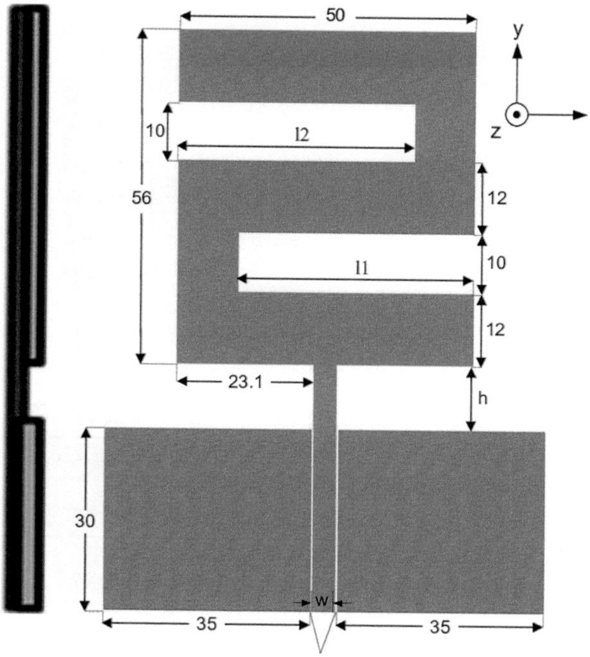

Figure 3.6 Z-shaped CPW-fed printed antenna configuration [16].

The simulating results of the Z-shaped antenna are done using a high-frequency structure simulator (HFSS) tool and the experimental done by a network analyzer. The experimental analysis of the return loss of the printed antenna from the network analyzer is shown in Fig. 3.8. Results have been measured outside the furnace, as the surface treatment characterizations remain the same for a period. Such a result illustrates that the design of the Z-shaped has a notch at 1.57 GHz with an amplitude of around −19 dB when the printed sample is exposed to a temperature equal to 180 °C. While the sample is exposed to a temperature less than 180 °C, there will be two notches at 780 MHz and 1.4 GHz with amplitude less than –19 dB, which reflects the effect of temperature on antenna performance.

Figure 3.7 Z-shaped printed antenna on glossy paper after annealing.

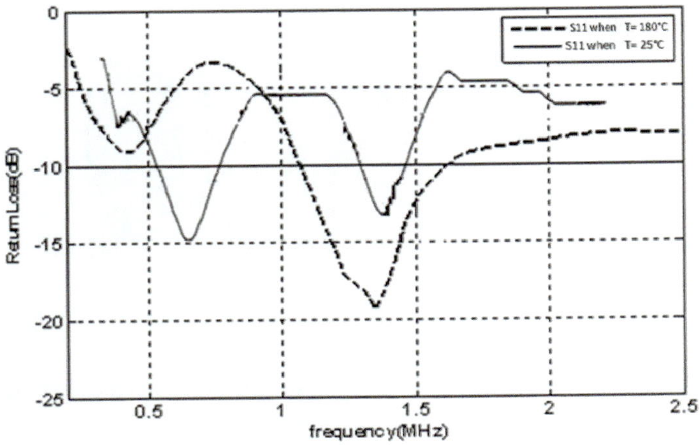

Figure 3.8 Measured return loss (S_{11}) of the printed Z-shaped antenna on glossy paper.

This is promising for fabricating a flexible and cheap antenna with such values, which depends on conductive silver nanoparticles contributing to the relatively high surface-to-volume ratio and sheet resistance. Antenna impedance is measured and equals 12.5 + j75 Ω. Figures 3.9 and 3.10 show the radiation patterns of the gain of the E-plane and H-plane, respectively. It can be noticed that there is a more directive pattern and high gain (around −8 dB) for the proposed flexible paper-based Z-shaped antenna.

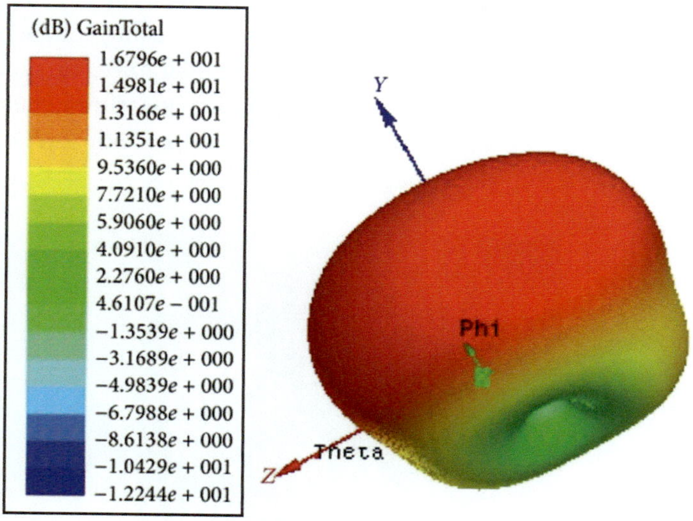

Figure 3.9 3D model of the radiation pattern of the printed Z-shaped antenna.

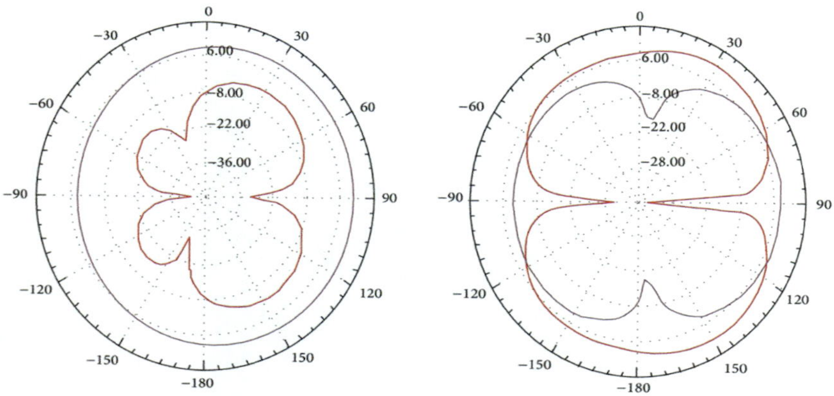

Figure 3.10 2D model of the radiation pattern of the gain of E-plane and H-plane.

3.5 Impact of Substrate on Flexible Low-Cost RFID Inkjet-Printed Antenna

This section investigates the effect of using different substrates on the performance metric of an antenna. The design of the Z-shaped antenna is printed on glossy paper and transparent PET films. Antennas are connected with an SMA connector using conductive silver epoxy with a ratio of 1:1 at room temperature, as shown in Figs. 3.11 and 3.12. Figure 3.13 presents measurements of return loss (S_{11}) using a network analyzer for both Z-shaped antennas printed on glossy paper and PET films.

Figure 3.11 Z-shaped antenna printed on PET films.

Figure 3.12 Z-shaped antenna printed on glossy paper.

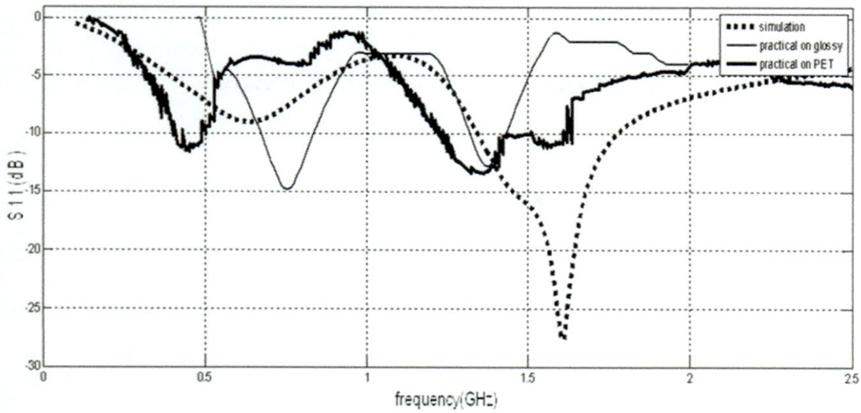

Figure 3.13 Return loss on network analyzer of the printed antenna.

It presents a different performance as compared to what has been mentioned in the last section, due to the effect of annealing. Figure 3.13 shows dual-band frequencies at 780 MHz and 1.4 GHz. Figures from 3.14 to 3.17 show the 3D and 2D models of the radiation patterns of the gain of E-plane and H-plane, respectively, for printed antennas on glossy paper and PET films. It can be noticed that the radiation pattern on PET is higher than the glossy paper.

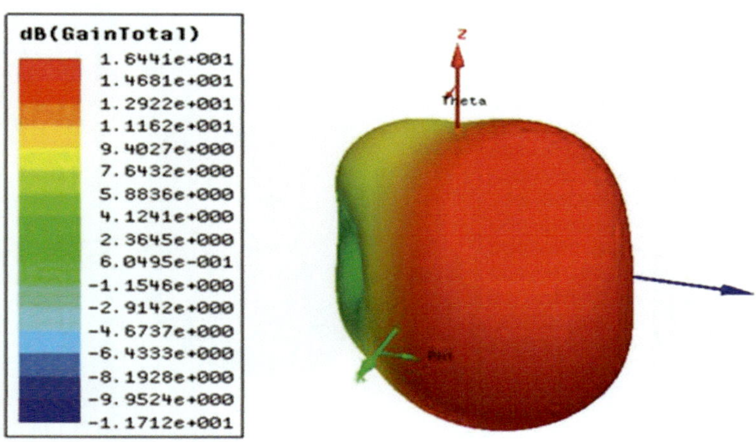

Figure 3.14 3D model of the gain for a Z-shaped antenna printed on glossy paper.

This section proposed a trail to get useful from different flexible substrate materials to tune the resonant frequency and radiation pattern. Therefore, a suitable selection of substrates will affect the performance metric of the designed system.

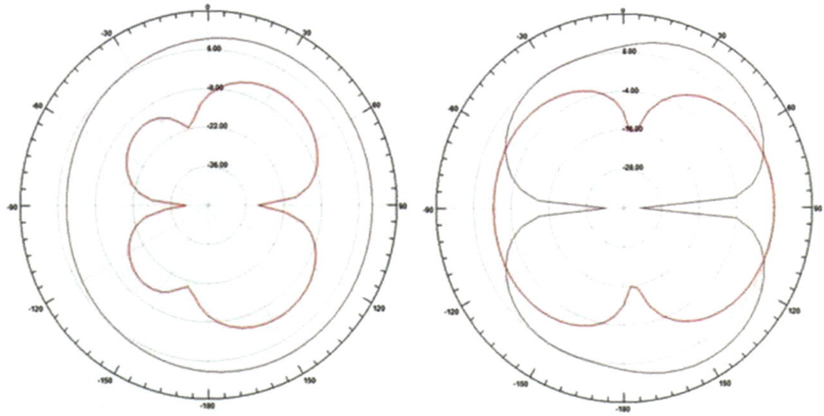

Figure 3.15 2D model of the gain of E-plane and H-plane for a Z-shaped antenna printed on glossy paper.

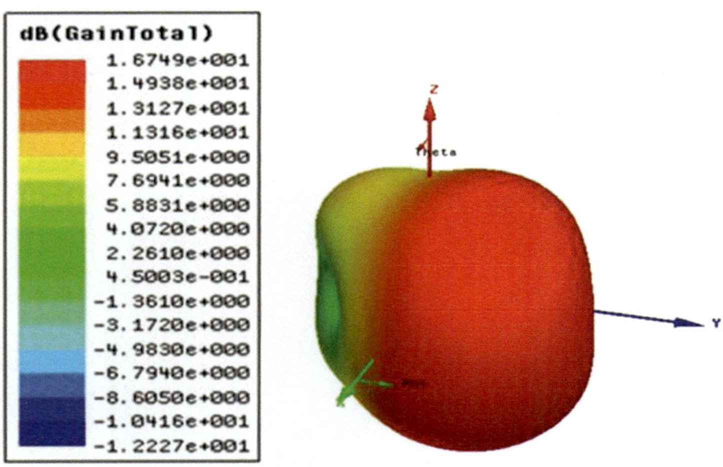

Figure 3.16 3D model of the gain for a Z-shaped antenna printed on PET films.

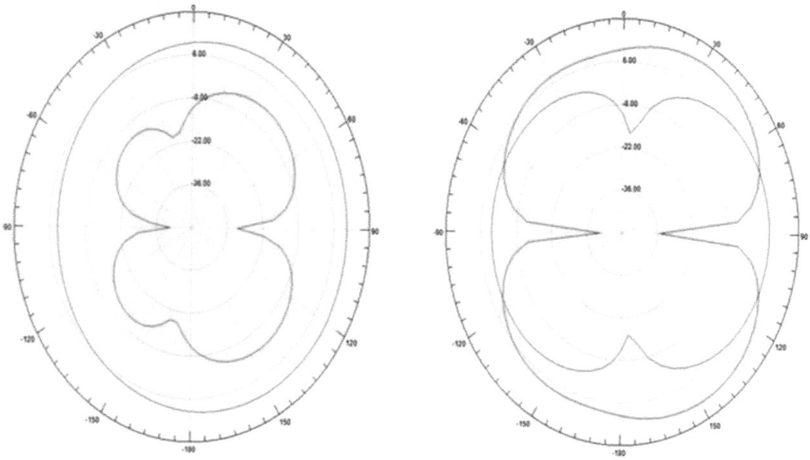

Figure 3.17 2D model of the gain of E-plane and H-plane for Z-shaped antenna printed on PET films.

3.6 Measurement Devices

There are several instruments that are used to determine if the antenna properties are optimally suited for the designed application. The antenna is precisely analyzed using antenna measurement systems that measure basic parameters such as return loss, input impedance, gain, S-parameters, radiation pattern, etc. This chapter presents a brief description of the most common instruments that are used for antenna measurements.

A network analyzer is used for measuring electrical network parameters. The basic architecture of a network analyzer involves a signal generator, receiver, and display, as shown in Fig. 3.18. It is commonly used to measure S-parameters because the reflection and transmission of electrical networks are easy to measure at high frequencies. Moreover, there are other parameters measured such as Y-parameters, Z-parameters, and h-parameters. A network analyzer is used to characterize two-port networks and it is mainly used to get a return loss curve and Smith Chart of the input impedance.

Figure 3.18 Network analyzer device.

An anechoic chamber is a chamber that is used to measure the gain, efficiency, and radiation pattern of antennas. Its walls, ceiling, and floor are lined with special electromagnetic wave-absorbing jagged triangle-shaped materials.

The jagged triangle shapes are designed so that what is reflected from them tends to spread in random directions, and what is added together from all the random reflections tends to add incoherently and is thus suppressed further. Figure 3.19 shows a picture of an anechoic chamber during the pattern measurement process.

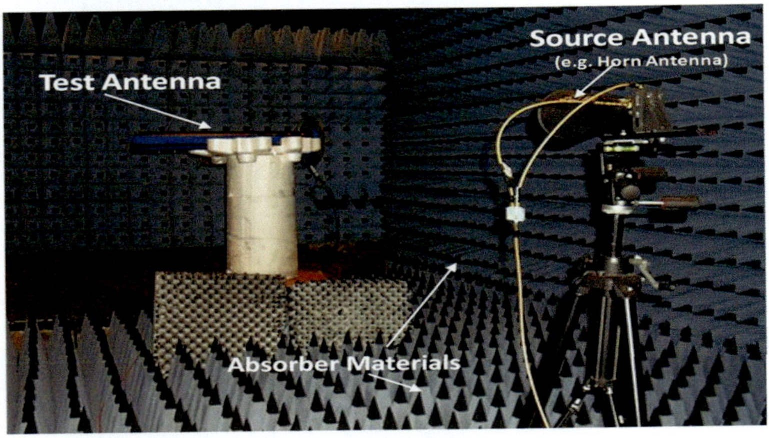

Figure 3.19 Example of an anechoic chamber [17].

3.7 Simulation Environment

There are a lot of parameters that should be considered in antenna design. These parameters will affect how much the design fits the requirements, as it is not only the dimension of the antenna (length (L) and width (W)), but there are more parameters that need to be considered such as [18]:

1. Substrate materials: FR-4, glossy, PET, PDMS, etc. and it presents dielectric constant (ε_r), loss tangent (δ), etc.
2. Substrate height (h)
3. Type of printed antenna material (copper, silver, etc.)
4. Dimension of radiation box
5. Matching impedance
6. Feeding position

These parameters and others play a significant role and affect the performance of the designed antennas. Therefore, needing for electromagnetic simulation tool can be used to consider these combinations of different parameters and give the ability to change their values to verify each design effectively. There are lots of software that are used to simulate the electromagnetic behavior of antenna designs. The most common simulators are Ansoft HFSS and computer simulation technology (CST).

This work focuses on the Ansoft HFSS software, because of its friendly user interface and ease of use and interface with other programs. It is also a full-wave electromagnetic (EM) software package for calculating the electromagnetic behavior of a 3D structure. HFSS is used for computing several parameters such as S-, Y-, Z-parameter matrix (2D plot), Smith Chart (port characteristic impedance), current distribution, E-field, H-field (3D static), and animated field plot in vector display or magnitude display. Far-field calculation in 2D and 3D for (gain, radiation pattern), material losses, radiation losses, and a lot of parameters can be measured [19].

HFSS gives users the ability to design and assign parameters of any antenna design shapes using the friendly user GUI interface, as shown in Fig. 3.20. There are five main different windows each is used to help in antenna designing. These windows are the 3D modeler area, project manager area, properties area, progress area, and message manager area. The function of each area is summarized, as follows:

- *3D Modeler Window*: This area is used to create the proposed geometry model. This window consists of the model view area (or grid) and the history tree. The history tree documents the actions that have been taken in the model view area and provides an alternative way to select objects in the model view area.
- *Project Manager Window*: This area is used to display details about all open HFSS projects. Each project ultimately includes a geometric model, its boundary conditions and material assignments, and field solution and post-processing information. It also has a lot of options used to assign material of substrate, antenna, etc., besides the ability to select the interested parameters for measurements after assigning sweep frequency and resolution step.

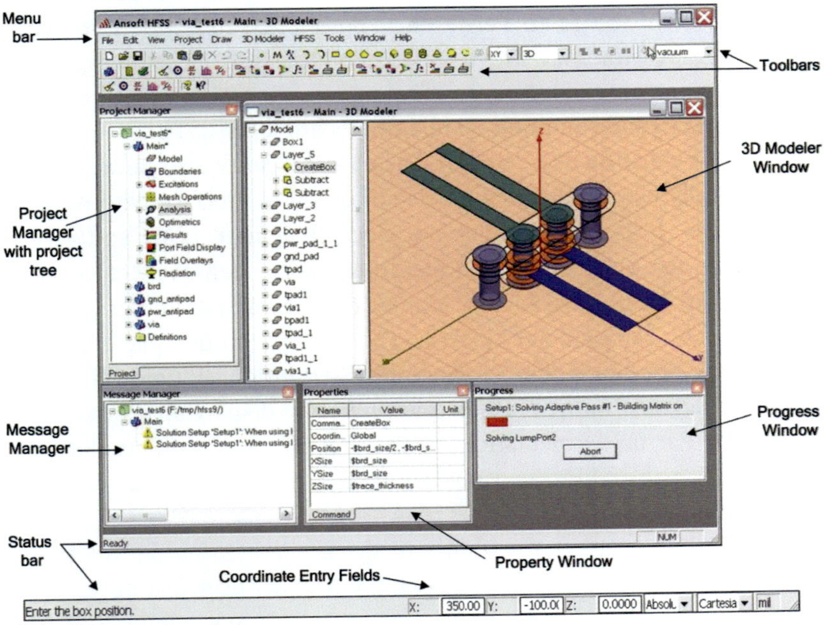

Figure 3.20 Illustration of desktop interface window of Ansoft HFSS [20].

- *Properties Window:* This area consists of two tabs: the command tab and the attribute tab. The command tab is used to display information about selected actions in the history tree that were performed to either create an object or modify an object. The attribute tab also displays information about the material and the properties of a selected object.
- *Progress Window:* This area is used when a simulation is running to monitor the solution's progress.
- *Message Manager Window:* This area is used to display messages associated with a project's development such as warning and error messages, execution progress, etc.

HFSS is based on Microsoft Visual Basic Scripting Edition (VB) scripting language to record macros. The usage of HFSS scripting is a fast, effective way to accomplish tasks that are repeated several times or in case you want to make some modifications and changes to antenna design; for example, change sweeping frequency values. HFSS gives the ability to run the written script (code) by any text editor after mapping it into VB format. Although HFSS records scripts in the VBS extension, it can also execute scripts in JavaScript format [19].

In this work, Ansoft HFSS is used for drawing antenna design and as-signing its various parameters using scripting and coding files. MATLAB programming language is used as a scripting language for designing and simulations. MATLAB-HFSS scripting helps in creating, running, and controlling the design on HFSS from the MATLAB window. This provides the ability to share results between MATLAB and HFSS, which is useful for adapting operating parameters over run time.

This method is efficient and massively used when it is required to change different antenna design parameters to achieve certain design criteria. For example, optimum operating parameters at a certain operating frequency with minimum return loss, high gain, etc. This is provided by a little bit of changing in design such as adding small gaps/slots or cutting on edges, etc. This method offers a fast and accurate technique that can be used to interface HFSS based on pre-assigned parameters and different operating scenarios coding on MATLAB. Figure 3.21 summarizes the interfacing close-loop diagram between MATLAB and HFSS.

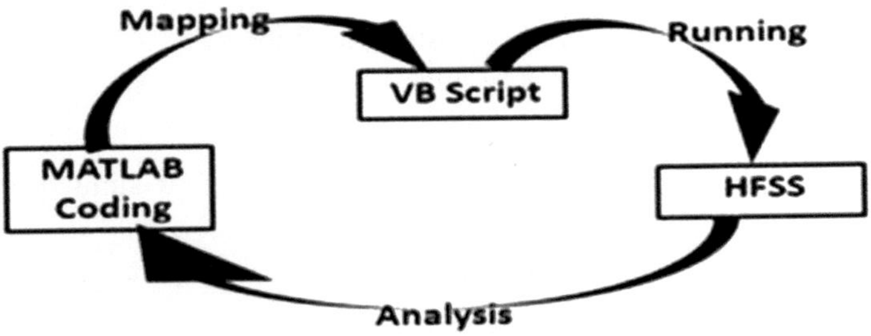

Figure 3.21 Illustration of basic operation of MATLAB-HFSS scripting closed loop.

It starts with mapping MATLAB codes to VB script format (.vbs) and then sends these designs for running on HFSS based on pre-assigned parameters from MATLAB. The results are calculated and saved and then returned to MATLAB for checking, analyzing, and displaying the required design output metric. The most common techniques that are useful from MATLAB-HFSS scripting are optimization algorithms such as genetic algorithm (GA) [20–23].

References

1. V. Bhatnagar and P. Owende, "Energy harvesting for assistive and mobile applications," *Energy Science & Engineering*, vol. 3, no. 3, pp. 153–173, 2015.
2. L. Yang, A. Rida, R. Vyas, and M. M. Tentzeris, "RFID tag and RF structures on a paper substrate using inkjet-printing technology," *IEEE Transactions on Microwave Theory and Techniques*, vol. 55, no. 12, pp. 2894–2901, Dec 2007.
3. Z. Konstas, Z. Konstas, A. Rida, R. Vyas, K. Katsibas, N. Uzunoglu, and M. M. Tentzeris, "A novel "green" inkjet-printed z-shaped monopole antenna for RFID applications," *EuCAP 2009*, 2009.
4. Y. Kawahara, S. Hodges, B. S. Cook, C. Zhang, and G. D. Abowd, "Instant inkjet circuits: Lab-based inkjet printing to support rapid prototyping of UbiComp devices," in *Proceedings of the 2013*

ACM International Joint Conference on Pervasive and Ubiquitou Computing. ACM, pp. 363–372, 2013.

5. Z. Konstas, A. Rida, R. Vyas, K. Katsibas, N. Uzunoglu, and M. M. Tentzeris, "A novel "green" inkjet-printed z-shaped monopole antenna for RFID applications," *EuCAP 2009*, 2009.
6. A. Mansour, N. Shehata, B. Hamza, and M. Rizk, "Efficient design of flexible and low cost paper-based inkjet-printed antenna," *International Journal of Antennas and Propagation*, Article ID 845042, 2015.
7. D. Betancourt and J. Castan, "Printed antenna on flexible low cost PET substrate for UHF applications," *Progress In Electromagnetics Research C*, vol. 38, pp. 129–140, 2013.
8. H. R. Khaleel, H. M. Al-Rizzo, and D. G. Rucker, "Compact polyimide-based antennas for flexible displays," *Journal of Display Technology*, vol. 8, no. 2, pp. 91–97, 2012.
9. A. C. Durgun, C. A. Balanis, C. R. Birtcher, and D. R. Allee, "Design, simulation, fabrication and testing of flexible bow-tie antennas," *IEEE Transactions on Antennas and Propagation*, vol. 59, no. 12, pp. 4425–4435, 2011.
10. L. Marnat and A. Shamim, "Liquid crystal polymer (LCP) based antenna for flexible system on package (SOP) applications," in *Antenna Technology and Applied Electromagnetics (ANTEM), 2012 15th International Symposium on*. IEEE, pp. 1–4, 2012.
11. A. Rida, L. Yang, R. Vyas, S. Bhattacharya, and M. M. Tentzeris, "Design and integration of inkjet-printed paper-based UHF components for RFID and ubiquitous sensing applications," in *Microwave Conference, 2007*. European. IEEE, pp. 724–727, 2007.
12. L. Yang, R. Zhang, D. Staiculescu, C. P. Wong, and M. M. Tentzeris, "A novel conformal RFID-enabled module utilizing inkjet-printed antennas and carbon nanotubes for gas-detection applications," *IEEE Antennas and Wireless Propagation Letters*, vol. 8, pp. 653–656, 2009.
13. D. M. Pozar, "Microstrip antennas," *Proceedings of the IEEE*, vol. 80, no. 1, pp. 79–91, 1992.
14. R. Mannerbro and M. Ranlöf, "Inkjet and screen printed electro-chemical organic electronics," 2007.
15. I. A. Grimaldi, "Realization and optimization of organic electronic devices manufactured by innovative technique: Inkjet printing," 2011.

16. Z. Konstas, Z. Konstas, A. Rida, R. Vyas, K. Katsibas, N. Uzunoglu, and M. M. Tentzeris, "A novel "green" inkjet-printed z-shaped monopole antenna for RFID applications," *EuCAP 2009*, 2009.

17. http://www.antennatheory.com/measurements/antenna.php (last access on 01/05/2017).

18. J. L. Young, "Advanced microwave ferrite research (AMFeR): Phase three," DTIC Document, Tech. Rep., 2008.

19. "Hfss v10.0 user manual," 2005.

20. S. Chakraborty and U. Mukherjee, "Micro-strip antenna optimization using genetic algorithms," in *Computer and Communication Technology (ICCCT), 2010 International Conference on*. IEEE, pp. 635–640, 2010.

21. A. Wyant and J. Venkataraman, "Optimization of reduced size microstrip patch with slots using a genetic algorithm," *Department of Electrical Engineering, Rochester Institute of Technology*, 2008.

22. S. Chouksey and M. Ghadle, "A genetic algorithm for optimization of microstrip patch antenna," *International Journal of Electrical, Electronics and Computer Engineering*, vol. 1, no. 2, pp. 97–99, 2012.

23. A. M. Mansour, B. Mokhtar, K. Gomah, K. Marghany, A. Abdelmonsef, M. R. M. Rizk, and N. Shehata, "Compact reconfigurable multi-size pixel antenna for cognitive radio networks and IoT environments," in *Antennas & Propagation Conference (LAPC), 2016 Loughborough*. IEEE., pp. 1–5, 2016.

Chapter 4

Flexible UWB Antennas on Polymer-Based Conductive Material

Thomas Peter and Rajagopal Nilavalan

Department of Electronic & Computer Engineering, Brunel University London, Uxbridge, Middlesex, UK

thomas@servcommintl.com

4.1 Introduction: An Explanation of the Process and Approach

During the recent decade, many new design concepts and techniques, materials, and fabrication processes have been introduced for the development of flexible antennas for consumer electronics. Researchers, both in industry and academia, are continuously relying on flexible antenna technology to solve the problems of efficiency, reliability, cost, weight, shape, and reproducibility. In this regard, the need to use environmentally friendly and cost-effective materials and production processes for such technology to be ubiquitously employed in next-generation communications is ever more evident.

Flexible Inkjet-Printed Antennas: Design, Fabrication, and Applications
Edited by Nader Shehata

ISBN 978-981-4968-87-4 (Hardcover), TBA (eBook)
www.jennystanford.com

DOI: 10.1201/9781003774754-4

This chapter aims to reflect on flexible antennas conceived on polymer-based conductive materials, specifically UWB antennas on transparent conductive oxide polymer films, AgHT-4, and AgHT-8. The chapter thus deals with the design, numerical simulation, fabrication process, techniques to improve efficiency, flexibility test, and measurements of the flexible UWB antenna. The chapter also introduces a self-assembling silver nano-particle technology (SANTE)-based flexible conductive polymer as an alternative material for next-generation flexible antennas. SANTE evolved as a spin-off of inkjet printing technology and has certain characteristics that can additionally enhance the properties of antennas produced using this technology.

4.2 Properties of the Transparent Conductive Oxide AgHT

Transparent conductive oxide (TCO) films can be prepared from a variety of materials including semiconducting oxides of tin, zinc, indium, gold, silver, and titanium [1]. A review of the physical properties of the TCO, AgHT is introduced in this section and compared, to gain an insight into understanding how it would affect the performance of an antenna made from this material.

A figure of merit for a TCO may be effectively defined as the ratio of its electrical conductivity to the optical absorption coefficient of the film. Besides the figure of merit, conductivity; physical, chemical, and thermal durability; etch-ability; work function; plasma wavelength; deposition temperature; toxicity; uniformity; and cost are other factors that may influence the selection of a particular TCO for an antenna [2].

4.3 Characteristics, Properties, and Applications of AgHT

The TCO AgHT has a variety of uses. One common use is its ability to reflect thermal infrared heat. This feature is exploited to make energy-conserving windows. For example, it is used use as a sun-shielding film, manufactured by Solutia Inc. [3]. This low-emissivity window application (Fig. 4.1) is currently the largest area of use for AgHT. Other applications include projector lamp shields, theatrical lighting, and electronic display filters.

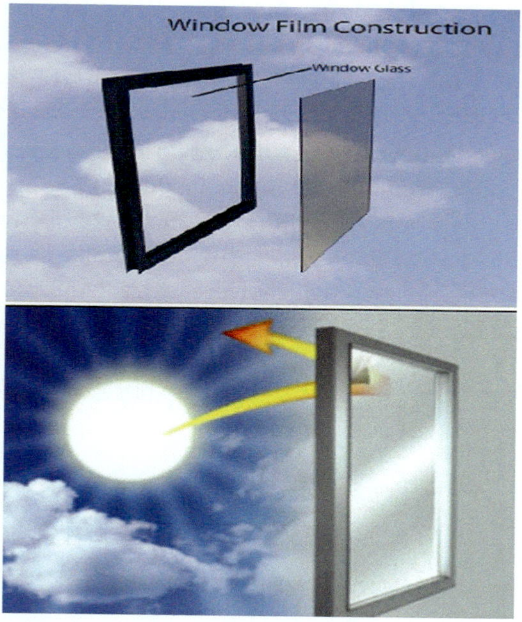

Figure 4.1 AgHT used for sun shielding from harmful UV rays [3].

However, the use of AgHT for ultra-wideband antennas other than narrow-band applications is still in the research stage basically because of the roller-coaster nature of the regulatory support for UWB technology, which had been received in the past and witnessed changes from commissioning bodies of Europe. With the first AgHT UWB antenna being demonstrated by Katsournaros et al. [6], more research is being done to improve the capability and performance of TCO-based UWBs.

AgHT was the TCO used to demonstrate the feasibility of the first transparent narrowband antenna in 1997 [4]. Ironically, AgHT was also the TCO used to demonstrate the first UWB antenna [5]. AgHT, i.e., AgHT-4 and AgHT-8, has been notably used for EMI/RFI shielding and infrared reflection applications in buildings, cars, and other applications. It is produced by a subsidiary of Solutia Inc., CP Films (formerly known as Courtaulds Performance Films). In recent times, Solutia has been bought over by Eastman Chemical Company, USA. CP Films produces the two trademark variants of the AgHT, i.e., AgHT-4 and AgHT-8, for the world market. AgHT is also produced by American Polarizers Incorporated (API), USA, and Instrument Plastics, UK. AgHT-4 has a surface resistance of 4 Ohm-sq while AgHT-8 has a surface resistance of 8 Ohm-sq; basically, the number designation is an indicator of the product's approximate surface resistance. 4 Ohm-sq is equivalent to a conductivity of 250,000 S/m and 8 Ohm-sq to a conductivity of 125,000 S/m since conductivity is the inverse of resistivity. Figure 4.2 provides details of AgHT's light transmittance properties, which were extracted from CP Films' AgHT datasheet.

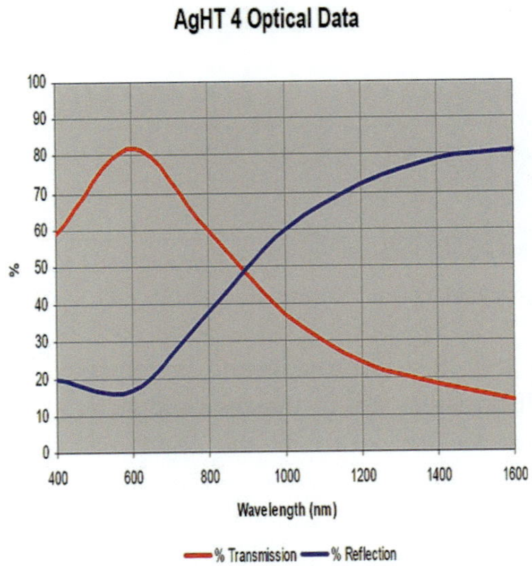

Properties of Standard AgHT™ Products

	AgHT-4	AgHT-8
Visible Transmittance	Min. 75%	Min. 82%
Infrared Reflectance	Min. 80%	Min. 75%
Surface Resistance (ohms per square)	4.5 ± 1.0	8.0 ± 2.0
Shielding Effectiveness*	24 - 44 dB	20 - 40 dB

Figure 4.2 Basic properties of AgHT [2].

As can be seen, AgHT has a visible light transmittance (VBLT) of between 75 and 82, which makes it quite suitable for use in conjunction with DSSC [6] for solar energy harvesting in green technology applications.

4.4 Other Properties

Properties other than those reflected in Fig. 4.2 have been observed through experimental data and observations in relation to its use as an antenna material.

4.4.1 Transparency

As described earlier, AgHT became a potential candidate as an antenna material, when engineers at NASA successfully demonstrated a patch antenna radiating at 2.3 GHz and 9.5 GHz in 1997 [4]. AgHT is made of a 3-layered conductive coating on a polymer base. This 3-layered coating on a transparent polyethylene terephthalate (PET) or polymer, comprises a silver layer sandwiched between two layers of tin oxide. The coating is electrically conductive and transparent with a VBLT of 75–82%. The AgHT film has an approximate thickness of 0.175 mm. About 40% of this thickness is the coating and the balance is PET.

The AgHT film is strong enough to stand alone to give support to a miniature antenna designed on it. For above 50 mm × 50 mm guideline, some additional support is needed. A raw PET without coating can effectively be added via transparent adhesive to provide additional support above the 50 mm × 50 mm guideline and still maintain its VBLT. The supported structure can still be conformal.

The transparent features of the AgHT provide the antenna with excellent discreetness for use in stealth and covert operations similar to the Fractal Antenna System's "Tranztenna" [7] as shown in Fig. 4.3.

The optical transparency of AgHT is inversely related to its electrical conductivity, which, in turn, depends on the mobility of free electrons in the material. In order to improve the optical transparency of the AgHT, the conductive coating, in this case of silver, has to be as thin as possible. Consequently, the number of free electrons is also decreased thus decreasing the conductivity of the material. Alternatively, it can be said that the resistivity of the AgHT increases as the free electron mobility decreases with the thinning of the silver layer.

Figure 4.3 The Tranztenna mounted on the side window of a military jeep.

4.4.2 Electrical Conductivity

The electrical properties of TCOs are determined by their sheet resistance and this in turn is related to the specific resistivity ρ and the film thickness d, i.e., sheet resistance = ρ/d. Electrical conductivity is the inverse of the resistivity of the TCO material. As earlier stated, the electrical conductivity of AgHT-4 and AgHT-8 is 250,000 S/m and 125,000 S/m, respectively. In this section, it will be seen how conductivity is dependent on material properties and for this reason, the silver-based AgHT is a very suitable candidate for antenna materials.

For modern-day applications like touch screens and high-resolution displays, it is critical for the TCO to be as thin as possible. This is because if the film is thick, the etched patterns for the displays in the TCOs could create height variations in the device. Hence, the topography has to be as thin as possible. which is only possible with a thin as possible TCO. The AgHT's conductive layer, for example, is as thin as 0.0525 mm. However, the thinness comes with a price, namely, a reduction in conductivity. The electrical conductivity depends on the free electrons available on the conductive layer of the TCO. The conductivity, σ increases with the products of the concentration of free electrons and the mobility. For the metallic layer of silver in AgHT, the free electron concentration is fixed by the structure and electronic properties of the metal. In semiconductors, however, the free electron concentration is determined by the maximum number of electronically active dopant atoms in the lattice structure. Hence, a logical approach would be to increase the conductivity by increasing the number of dopant atoms; however, such attempts to place a large number of dopant atoms into the lattice produce a neutral defect that adversely affects or decreases the mobility and consequently the conductivity [2]. Table 4.1 attempts to summarize the order of electron concentration in terms of conductivity as it is generally found in TCOs.

Table 4.1 Order of conductivity (electron concentration) of TCOs

TCO type	Conductivity
Ag-based	↓ Decrease
TiN	
In_2O_3:Sn	
ZnO:Al	
SnO_2:F	
ZnO	

4.4.3 Plasma Frequency

The plasma frequency for the conduction electrons in a TCO can be considered as the dividing line or plasma edge for optical properties. This dividing line can be explained using the classical Drude free electron theory. The Drude model indicates that transmittance drops in the near-infrared region are associated with plasma frequency, *Wp*. At longer wavelengths (i.e., lower frequencies), reflection occurs because of the plasma edge, and light cannot be transmitted due to this phenomenon.

As such, at frequencies higher than the plasma frequency, the electrons cannot respond, and the TCO behaves as a dielectric. At frequencies below the plasma frequency, the TCO is characterized by high reflectance and functions as a screen of incident electromagnetic waves. For most TCOs, the plasma frequency is around the near-infrared part of the spectrum. As such, in this region, there is high reflectance, a feature that is used in AgHT to provide infrared screening in window applications. The visible or transmittance region in contrast is in the higher frequency range of the spectrum.

The plasma frequency increases approximately with the square root of the conduction-electron concentration. The maximum obtainable electron concentration and plasma frequency of TCOs in a general increase in the same order as resistivity. In essence, as the plasma wavelength decreases, the resistivity decreases resulting in an increased conductivity.

4.4.4 Deposition Temperature

This section explains why ITO and silver-based AgHT are the common commercially found polymer TCOs. In the manufacturing process of TCOs, a sufficiently high temperature needs to be maintained in order to develop the required properties in the TCOs. The list in Table 4.2 gives the order in which the temperature usually increases with some TCOs. The list is given in descending order, i.e., from the highest temperature required to that of the lowest.

As such, it can be seen why silver or ITO is preferred for deposition on thermally sensitive substrates like polymer plastics. Cadmium Stannate (Cd_2SnO_4), because of its high-temperature requirement, may require very refractory substrates to develop its best properties [2].

Also, it is found that ITO is commercially available on polymer plastics like PET only at a resistivity of 50 Ohm-sq and above. This is the lowest available in the market unlike AgHT, which is available at 4 Ohm-sq. ITO becomes too brittle and is susceptible to electrical discontinuity on polymers if subject to bending during handling or otherwise. As such, ITO at 4 Ohm-sq can only be found on rigid bases with glass being the most common.

Table 4.2 Deposition temperature of TCOs

TCO type	Deposition temperature
Cd_2SnO_4	Highest
SnO_2	↓
ZnO	↓
ITO	↓
Ag	Lowest

On the other hand, the silver-based AgHT, through a patented manufacturing process and design, is able to overcome this and maintain its electrical conductivity when flexed or bent; however, with some limits.

4.4.5 Durability and Etching Patterns on AgHT

Durability in the TCO environment is its ability to withstand a corrosive chemical environment, which is inversely related to its ease of etching. Tin oxide is the most resistant and as such its application in AgHT. Silver is easily tarnished by air and its moisture. Hence when handling AgHT film, it is advised to use vinyl gloves to avoid sweat from palms and fingers getting onto the surface of the AgHT film. The film normally comes with a layer of barrier film and it is best to leave the barrier film intact on the surface of the film if the geometry of the antenna is to be manually inscribed with a scalpel.

If the etching method is to be used, e.g., for displays, heaters, or even antennas, parts that need to be removed are done so using etchants. A list [2] of which is shown in Table 4.3. From the table, which is listed with materials with the highest ease of etching at the top, it is noticed that zinc oxide is the easiest and tin oxide the most difficult.

As it is not possible to carry out etching on an already manufactured AgHT film, and as mentioned earlier, tin oxide is quite resistant to etching; alternative approaches have been used to fabricate AgHT antennas. Antenna geometry using AgHT is normally "cut out" in papers published by other researchers. However, the author has introduced an inscribing method in his later papers [8–10] to minimize the difficulty of attaining narrow feed gaps. It is also possible to develop an "imprint-cut technology" to imprint the geometry of the antenna through very sharp-edged dies onto the surface of the AgHT polymer as an improvement of this approach to attain a precise and repeatable prototype. AgHT as mentioned earlier is comprised of a thin layer of silver sandwiched between two layers of tin oxide coated onto a film of PET.

A milling method that will not damage the coating surface can be implemented. For this method, a high-precision milling machine will be required especially one that is capable of milling slots of 0.05–0.1 mm in width.

Table 4.3 Etchants for TCOs

TCO type	Etchants
ZnO	Dilute acids
ZnO	Ammonium chloride
TiN	$H_2O + NH_3$
In_2O_3	$HCl + HNO_3$ or $FeCl_3$
SnO_2	Zn + HCl
SnO_2	$CrCl_2$

Laser cutting is not suitable as the heat generated at the cutting point causes discoloration along the edges of the geometry that is traced or cut off as well as damage to the conductive material along those edges. This will greatly affect the performance of the antenna that is made in this way.

4.4.6 Hardness of AgHT

The mechanical durability of TCOs is related to the hardness of the crystals from which they are formed [2]. Hardness values are usually given in Mohs where higher values indicate harder materials.

Tin oxide is known to be even harder than glass, hence its popularity in the manufacture of TCOs especially, in applications that are exposed to contact. Zinc oxide is one of the easiest to be scratched. Thin silver films are very fragile and need to be coated with protective layers just like in AgHT, where the protection (Fig. 4.4) is provided by the tin oxide.

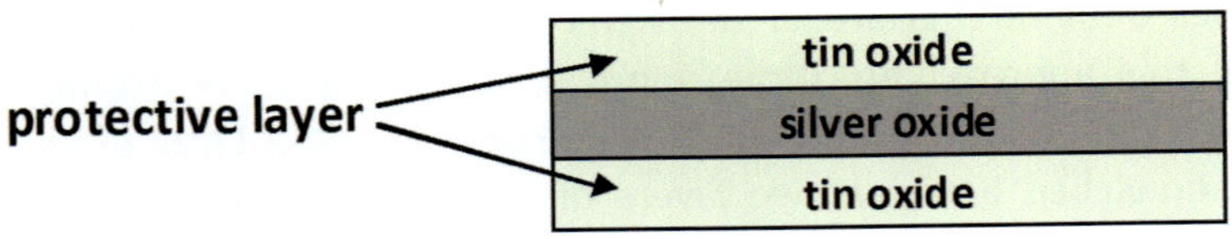

Figure 4.4 Composite structure of the AgHT film.

4.4.7 Sun Shielding and Low Emissivity

As already shown in Fig. 4.4, one of the most common uses of AgHT is sunscreens or low emissivity films on window glass. Integrating antennas onto AgHT films is merely taking advantage of its conductive properties to provide multiple applications using a single material.

AgHT when applied on glass windows improves the energy efficiency of the windows because free electrons reflect infrared radiation for wavelengths longer than the plasma wavelength. During winter or in cold climates, the plasma wavelength should be fairly long about 2 μm so that most of the solar spectrum is transmitted into heat inside the building [2]. For hot climates, conversely a short plasma wavelength, ≤1 μm is desirable so that the near-infrared portion of the incident light can be reflected out of the building.

Metals such as silver have short plasma wavelengths, hence their use in AgHT for low emissivity windows. Normally, the silver layer because of its poor durability is sealed inside double-glazed panes for protection from air and moisture. This protection is already provided in AgHT as the silver layer is sandwiched, as shown earlier in Fig. 4.4, between two layers of durable tin oxide.

4.5 Conformability and Flexibility of AgHT Antenna

AgHT can be flexed or conformed to the shape of the mount like circular columns, right-angle corners, or the edge of a laptop. However, there is a slight limitation to this with regard to the smallest antenna size that can be confirmed. Experimentally, it is possible to bend antennas with a width of 30 mm and above. Below that because of the ground connection to the connector legs, for example, in CPW feeds, makes it difficult to bend. This may be overcome through the use of cable wire connections. Also, a 90-degree bending of a small-width antenna might introduce a break in the silver layer creating a disconnect in conductivity. In such cases, a 120-degree bend angle is the maximum suggested to avoid any disconnect as illustrated in Fig. 4.5.

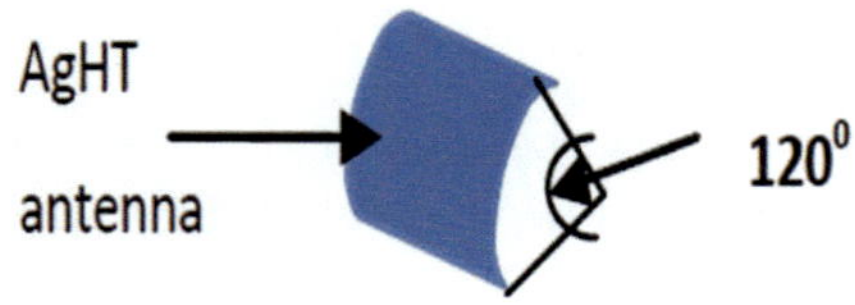

Figure 4.5 Maximum bend angle recommended for smaller antennas.

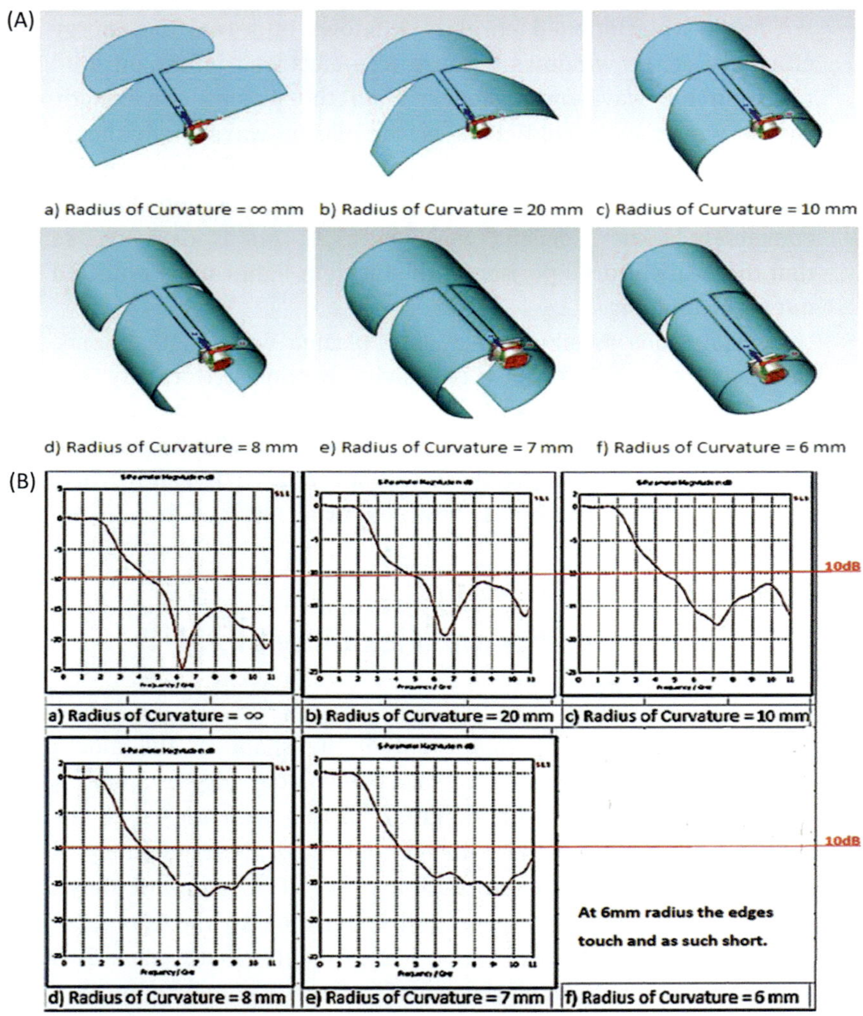

Figure 4.6 Antenna design and S_{11} when being flexed through various radii of curvatures (A) AgHT UWB antenna being flexed; (B) return loss (S_{11}) of the antenna.

The effect on the performance of a mushroom-top AgHT-8 UWB antenna during bending or conforming it through various radii of curvature was studied in simulation using the full-wave simulation software, CST Microwave Studio. The simulated (S_{11}) seemed to improve rather than deteriorate as the radius or bend angle gets smaller as depicted in Fig. 4.6.

The simulated AgHT-8 UWB antenna delivered a stable performance throughout its flexure, i.e., it was able to maintain its return loss below the 10-dB line for various radii of curvature. This makes it a desirable feature for conformal antennas and highlights the potential of AgHT-8 being a good candidate for overcoming performance deterioration during bending in body motion for wearable antennas.

AgHT-8 is selected to demonstrate the design and fabrication of a flexible 50 Ω, co-planar fed waveguide (CPW) UWB antenna prototype. The AgHT-8 film is 0.175 mm thick and has a surface resistance of 8 Ohm-sq, which is equivalent to a conductivity of 1.25×10^5 S/m [11]. The AgHT-8 film coating is made up of a transparent silver layer sandwiched between two layers of tin oxide. The antenna is mounted on a 0.175-mm-thick PET substrate with a relative permittivity of 3.24.

Table 4.4 Dimensions of the antenna structure in Fig. 4.7a

Parameters	L	W	Y	L_1	L_2	W_2
Size (mm)	50	60	48	20	11	28.4

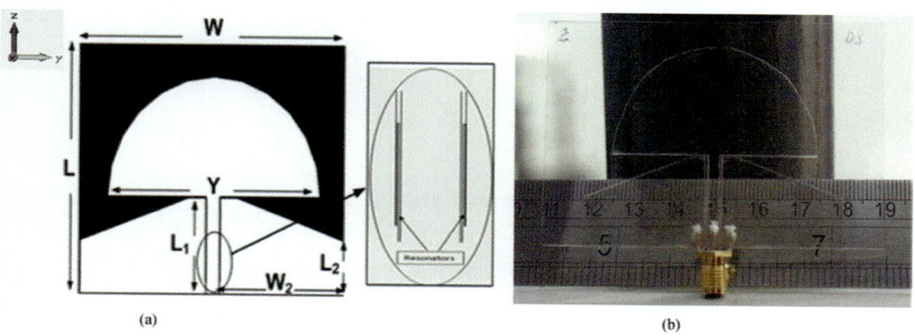

Figure 4.7 AgHT-8 UWB antenna: (a) AgHT-8 UWB antenna with resonators for impedance matching, (b) the designed optically transparent AgHT-8 UWB antenna.

The antenna is designed and optimized as a semicircular disc (mushroom-shaped) for size reduction and fed by a 50-CPW line as illustrated in Fig. 4.7a. The radiating semicircular monopole has a radius of 24 mm. The current in simulation is found to be mainly distributed along the edge of the semicircle monopole showing that, as already established in [12], the first resonant frequency is associated with the radiating semicircular monopole with the ultra-wideband characteristic being caused by the overlapping of multiple resonant harmonics.

The CPW feed line has a width of 3 mm and a spacing gap of 0.1 mm. The total dimensions of the antenna including the PET substrate are 50 mm × 60 mm × 0.350 mm. The AgHT-8 UWB antenna as such is compact and optically transparent as shown in Fig. 4.7b. The CPW grounds each have a width of 28.4 mm, giving a total width (*W*) of 60 mm from edge to edge including the CPW feed line and the spacing gap. The gap between the CPW grounds and the bottom edge of the radiating semicircular monopole is 0.03 mm. The ground planes are beveled [13] and have slots cut out on both walls of the ground planes along the strip gap to perform as resonators [14] to improve impedance matching and tune the lower and upper limits of the bandwidth.

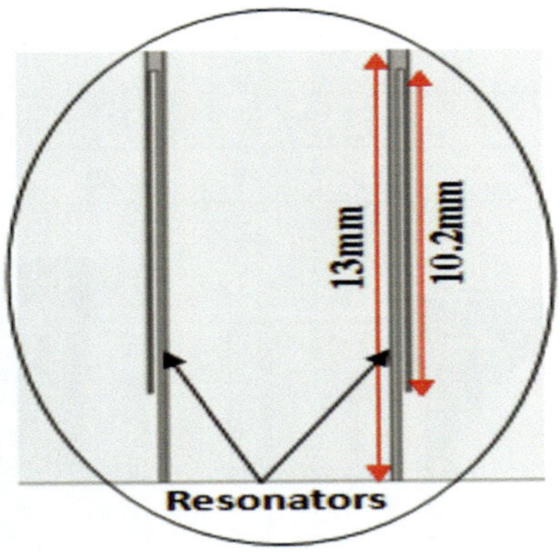

Figure 4.8 Resonator dimensions.

Impedance matching through the resonators is achieved by varying their heights. Increasing or reducing the height of the resonators gave the flexibility of tuning the bandwidth to obtain the optimized bandwidth range. A parametric study was done to obtain the optimum height of the resonators; the external height opposite the feed line was 13 mm whilst the shorter internal height of the resonator stub was 10.2 mm for the best impedance match as depicted in Fig. 4.8.

4.6 Improving Feed-line Conductivity Using a Silver Base

Normally, the connection of the feed points to polymer-based antennas is by thermal soldering of copper pads to the coaxial feeds or legs of the SMA connectors and gluing them to the surface of the polymer [11]. This approach has been used to avoid excessive thermal excursion, which could otherwise damage the polymer surface of the antenna, resulting in a poor connection or rendering the antenna useless. However, when applied to the surface of the AgHT-8 polymer's conductive coating, this copper pad technique results in additional reflection losses that may be further aggravated if the adhesive used to glue the pads is a too-thick layer or provides some electrical resistance itself. Cold soldering or non-thermal soldering using the electrically conductive paste is a much better option compared to thermal soldering as it allows direct soldering of the feed points to the antenna surface without damaging it. Although a better alternative, directly cold-soldering the coaxial feeds may also result in poor performance.

This is because the AgHT-8 coating is a three-layered coating made up of a layer of silver sandwiched between two layers of tin oxide. Direct cold soldering will only bring the coaxial feeds in contact with the topmost layer of the coating, the tin oxide layer, which has a lower electrical conductivity compared to that of the silver layer. As such, the tin oxide surface would become the main surface along which the electromagnetic waves propagate, resulting in poorer performance than if the feeds were connected to the silver layer [2]. Additionally, the smooth surface of the AgHT-8 material may also result in a dry joint making the bond weak. The apparent connectivity issues in these two techniques prevent maximum power transfer of the electromagnetic energy from the coaxial feeds besides the usual power transfer loss through poor impedance matching, resulting in overall poor antenna performance.

To overcome the above problems, a novel technique and a non-thermal soldering approach are introduced to improve the feed line conductivity. A silver base is applied as an interface to the silver layer of the AgHT-8 to improve the conducting surface for connectors. This results in a better antenna performance as better power transfer of the electromagnetic energy is achieved. In this approach, the coaxial cable is connected to the feed points by soldering an SMA connector using electrically conductive epoxy, but after removing the uppermost coating layer of the AgHT-8 film at areas on the CPW grounds and feed line where the connector legs of the SMA connector would sit, and applying a coat of silver paint on the exposed areas before actually soldering the SMA connector legs, as shown in Fig. 4.9. It is important to test for conductivity between the silver-coated areas and the feed line and CPW grounds before actually soldering the connector. This can be done by using a simple multi-meter to check for electrical continuity.

Figure 4.9 (Left) Antenna with the coating removed. (Right) Antenna after application of the conductive silver paint.

For comparison, three other antennas were similarly designed, but were mounted via:

1. Glued copper pads at the feed points as in [11].
2. Cold-soldered copper pads.
3. Direct soldering without a silver base at the SMA connector legs shows that improving the conductivity at the feed point will improve the performance of the antenna.

4.7 Fabrication and Measurement

The antennas were designed and simulated on CST Microwave Studio. The performance of the proposed approach was verified by fabricating the monopole UWB CPW-fed antenna and measuring the return loss. The measured return losses of four fabricated antennas are comparatively shown with the simulated result of the designed antenna with a 50 Ω port, as shown in Fig. 4.10. The return loss pattern of the silver-based antenna is the most similar to that of the simulated return loss, demonstrating that the proposed approach enables us to achieve a result closest to the ideal or designed performance.

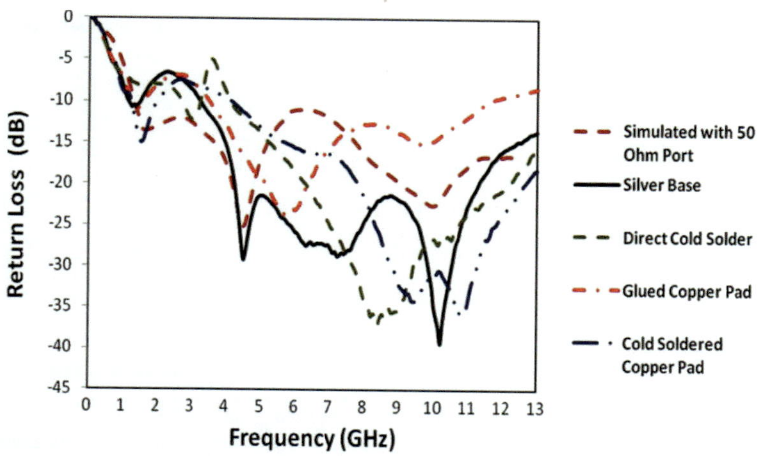

Figure 4.10 Measured return loss of the antennas versus simulated return loss with a 50 Ω port.

The silver-based antenna also showed comparatively better return loss performance than the one without. The return loss for the antenna fed via cold-soldered copper pads showed better performance than the one that had glued copper pads. This was probably due to the adhesive that was used to glue the copper pads to the antenna; it could have contributed to some electrical resistance. It was also observed that the compactness of the antenna gave it rigidity for independent mounting without any additional support. Of the four antennas, the antenna with the silver base was found to have a stronger bond by virtue of the entrenched sub-layer foundation of the silver base. This made the silver-based antenna a better candidate of the four antennas.

The measured efficiencies of the four prototype antennas are shown in Fig. 4.11, while the peak gains of the antennas are shown in Fig. 4.12.

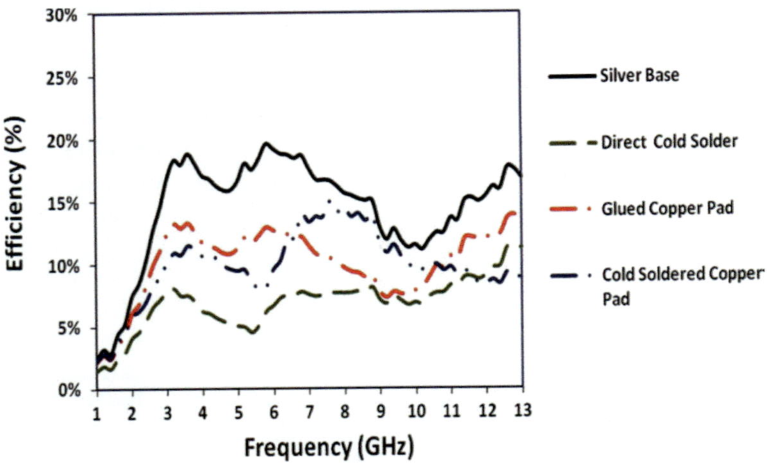

Figure 4.11 Measured efficiency of the four antennas.

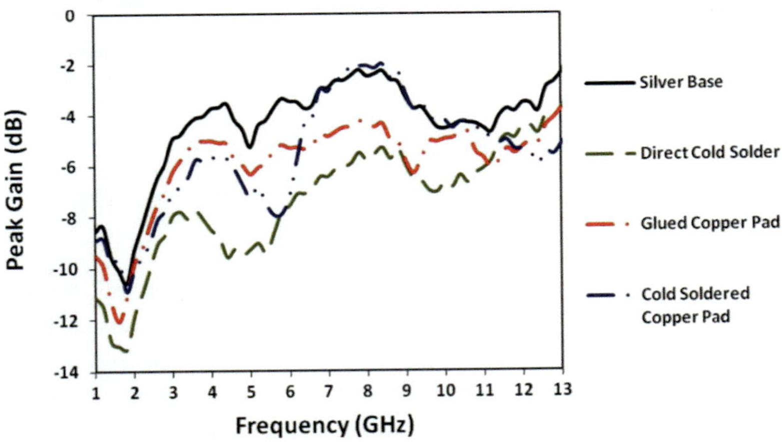

Figure 4.12 Measured peak gain of the four antennas.

The radiation patterns at 3 GHz and 7 GHz in the vertical plane and horizontal plane are shown in Fig. 4.13 and Fig. 4.14, respectively. The silver-based antenna demonstrates a smaller cross-polar pattern in the vertical plane compared to other antennas besides also demonstrating an omnidirectional characteristic in the horizontal plane.

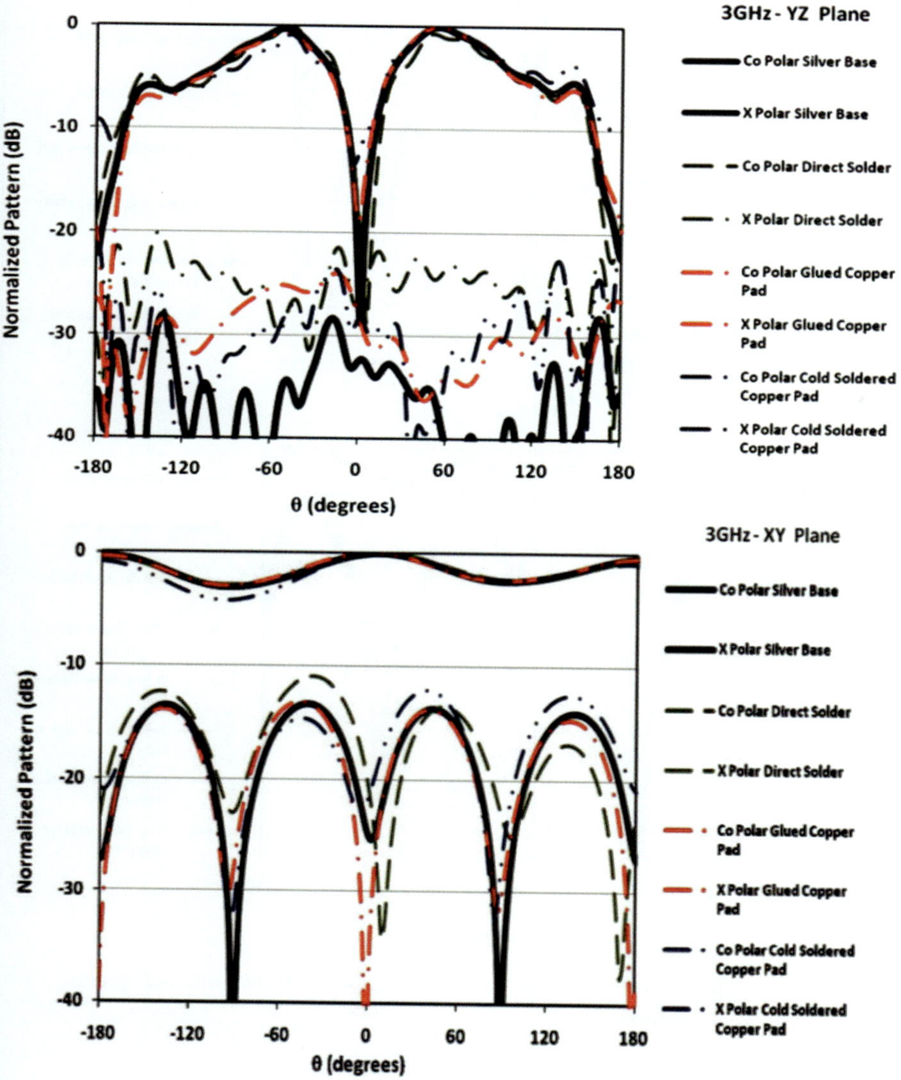

Figure 4.13 Measured radiation patterns in the YZ and XY planes for the four.

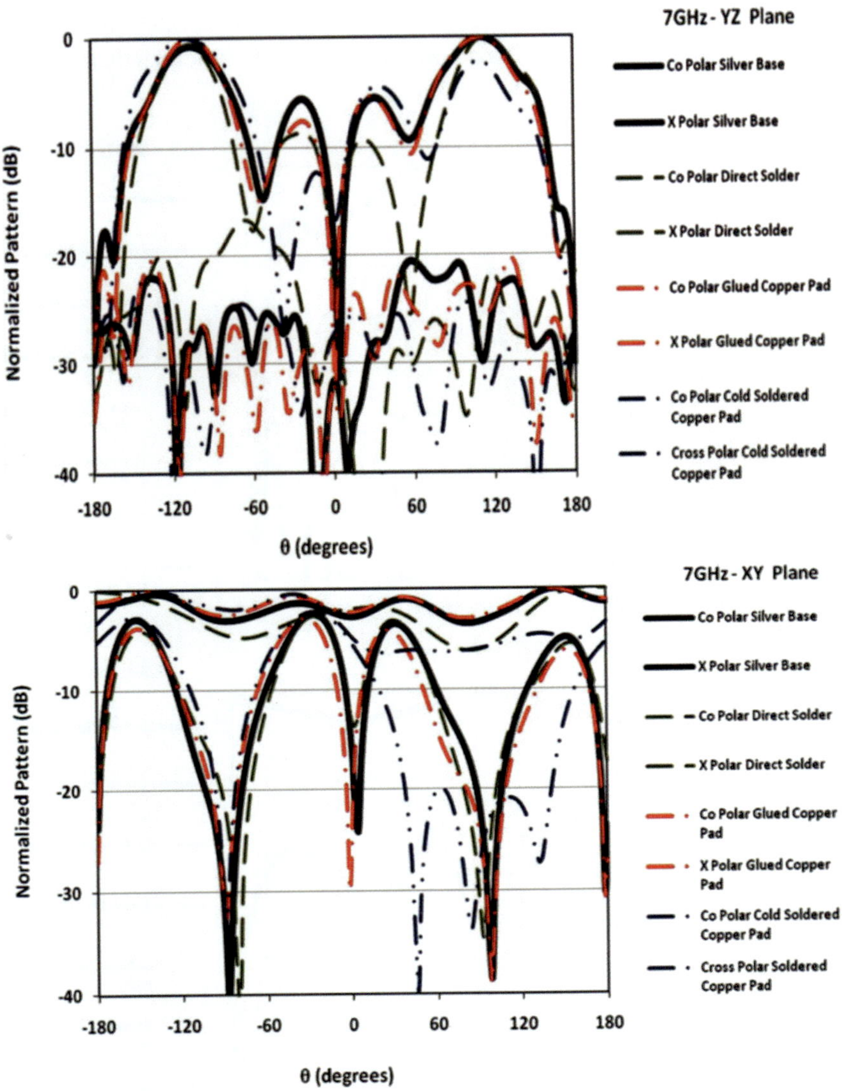

Figure 4.14 Measured radiation patterns in the YZ and XY planes for the four antennas.

4.8 New Technologies for Fabrication of Flexible Antennas on Polymers

In recent years, inkjet printing of RF circuits and antennas using highly conductive inks has become popular. New inkjet material printers operate by depositing ink droplets of size down to a few pico liters, and hence, these high-resolution printers can produce compact designs with tiny details accurately [14, 15]. This new technology utilizes conductive inks based on different nano-structural materials such as silver nano-particle-based ink, which is widely used due to its high conductivity. Evolving from this inkjet printing technology is a self-assembling SANTE, which can be applied to flexible polymers for next-generation flexible antennas.

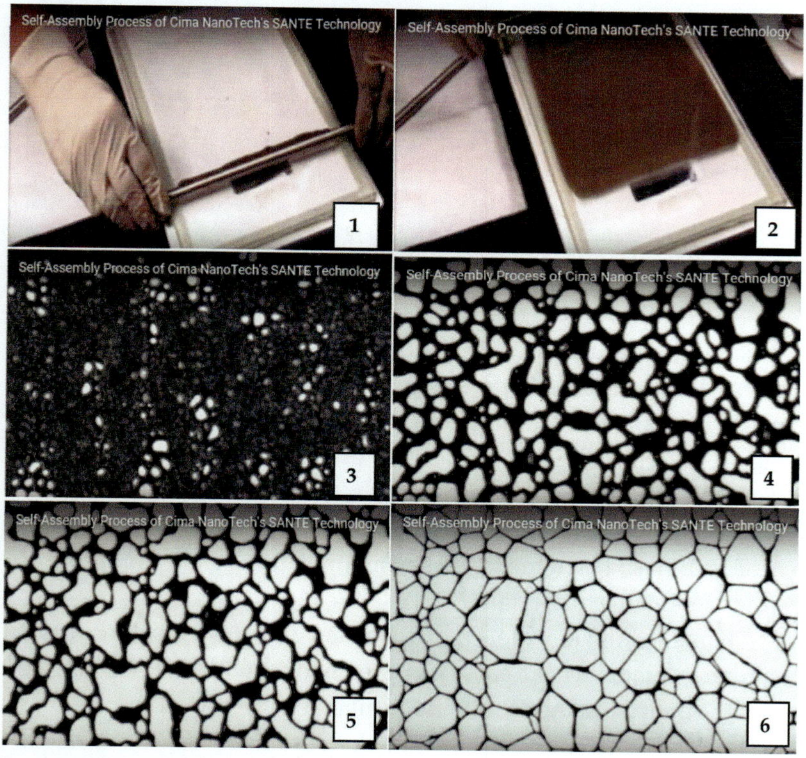

Figure 4.15 Step-by-step illustration of the self-assembly process of Cima NanoTech's SANTE Technology.

This patented technology by Cima NanoTech is a silver nanoparticle conductive coating that self-assembles into a random mesh-like network as it dries (Fig. 4.15), enabling products with excellent electrical conductivity for applications such as large projected capacitive touch screens, transparent antennas for internet of things (IoT), transparent heaters, and EMI shielding.

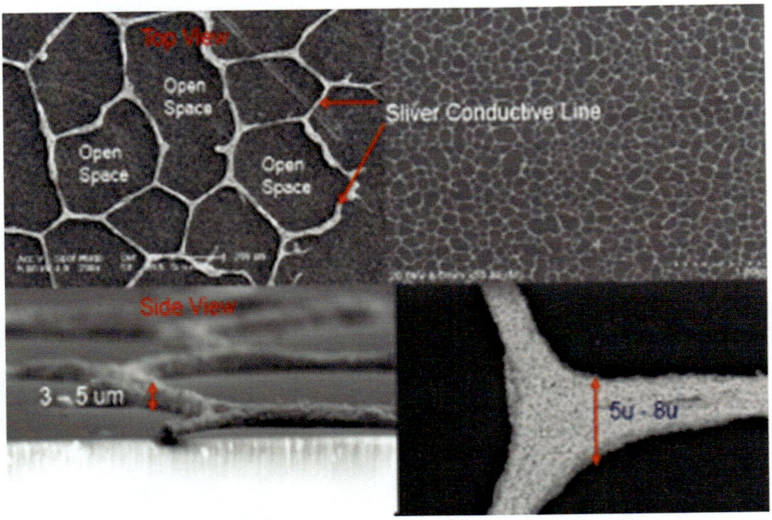

Figure 4.16 Microscopic view of the SANTE film structure.

In SANTE, when the silver particles assemble, they form a wire mesh with mesh widths and pores engineered to the desired sizes. A microscopic view of the material's structure shows the width and pore sizes of the silver wire mesh, as presented in Fig. 4.16. The silver wire meshes are additionally coated with layers of other conductive materials to meet the required level of conductivity or sheet resistance. As such, the conductivity of a SANTE-based polymer can be manipulated by the addition of other conductive materials.

Using SANTE, the performance of the flexible UWB antenna demonstrated earlier in the chapter with AgHT-8 can be greatly improved. AgHT is only available at resistivities of 4 and 8 Ohm/sq; whereas with SANTE, the resistivity can be brought down to even 0.05 Ohm/sq (currently available commercially), which is equivalent to a conductivity of 2,000,000 S/m, i.e., ×80 that of AgHT-4 and ×160 that of AgHT-8. Moreover, it is not as lossy as AgHT.

The improvement in performance of the mushroom top UWB antenna using SANTE is demonstrated in simulation and the simulated results are presented in Fig. 4.17 and Fig. 4.18.

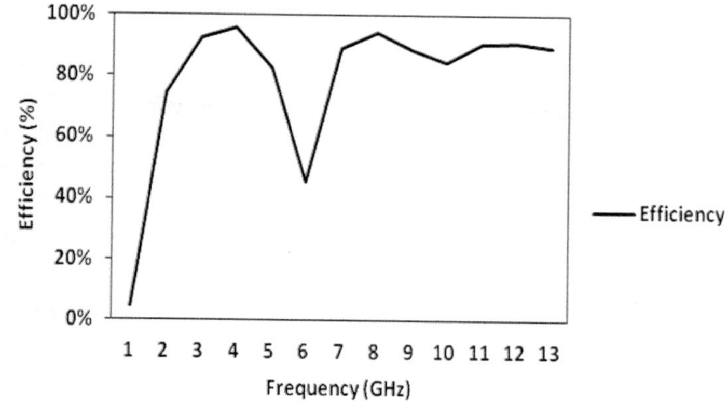

Figure 4.17 Simulated efficiency of the SANTE-based antenna.

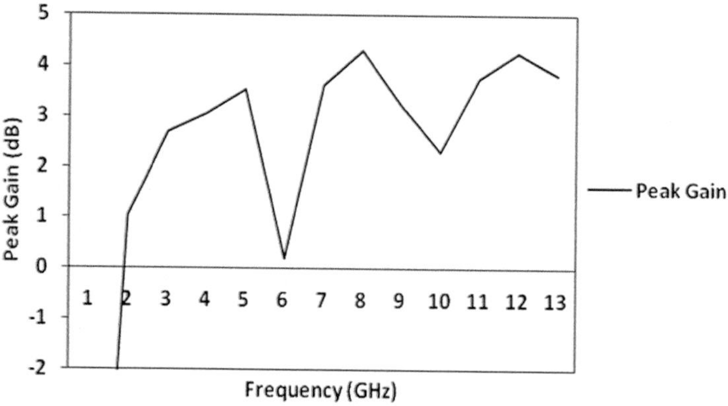

Figure 4.18 Simulated peak gain of the SANTE-based antenna.

It should be noted, however, that the gain and efficiency will differ in measurement and actually be lesser than in simulation depending on the total area of voids in the mesh on the radiating surface (or simply the effective area of the radiating surface) of the antenna. Though prototype antennas using this SANTE technology are not demonstrated in this chapter, the good performance of this SANTE-based material has been demonstrated in the author's co-authored papers on transparent branch-line couplers and phase shifters [16–18].

4.9 Conclusion

In this chapter, flexible antennas conceived on polymer-based conductive materials are discussed and presented in terms of design, fabrication, and measurement using a flexible and transparent AgHT-8-based UWB antenna. A spin-off of the technology of inkjet printing, SANTE, which can be used to produce flexible antennas has good performance as well as suitable for discrete applications. Such SANTE-based antennas can easily be integrated into flexible electronics for the internet of things (IoT) in next-generation communications.

References

1. R. G. Gordon, "Thin films for photovoltaic and related device applications," edited by D. Ginley, A. Catalano, H. W. Schock, C. Eberspacher, T. M. Peterson and T. Wada, *Mater. Res. Sco. Symp. Proc. 426, Pittsburgh*, 1996, p. 419.
2. R. G. Gordon, "Criteria for choosing transparent conductors," *MRS Bulletin*, vol. 25, pp. 52–57, 2000.
3. Solutia Inc. [Online]. Available: http://www.solutia.com/en/SolutiaCPFilmsVideo.aspx
4. R. N. Simons and R. Q. Lee, "Feasibility study of optically transparent microstrip patch antenna" *International Symposium and Radio Science Meeting cosponsored by IEEE, AP-S, and U.R.S.I.*, Montreal, Canada, July 13–18, 1997.
5. A. Katsounaros, Y. Hao, N. Collings and W. A. Crossland, "Optically transparent ultra-wideband antenna," *Electronics Letters*, vol. 45, no. 14, pp. 722–723, 2009.
6. DSSC. [Online]. Available: http://www.solarprint.ie.
7. Fractal Antenna Systems. Available: http://www.fractenna.com/whats/060228.html.
8. H. Gueddaoui, S. Maabed, G. Schmerber, M. Guemmaz and J. C. Parlebas, "Structural and optical properties of vanadium and hafnium nitride nanoscale films: Effect of stoichiometry," *The European Physical Journal B*, vol. 60, pp. 305–312, 2007.
9. T. Peter, R. Nilavalan and S. W. Cheung, "A novel transparent TSA for laptop and UWB applications," *PIERS Proceedings*, pp. 836–838, 2012.

10. T. Peter, T. A. Rahman, S. W. Cheung, R. Nilavalan, H. F. Abu Tarboush and A. Vilches, "A novel transparent UWB antenna for photovoltaic solar panel integration and RF energy harvesting," *IEEE Transaction on Antennas and Propagation*, vol. 62, no. 4, pp. 1844–1853, 2014.

11. J. S. Hyok, Y. H. Tsung, D. F. Sievenpiper, P. H. Hui, J. Schaffner and E. Yasan, "A method for improving the efficiency of transparent film antennas," *IEEE Antennas Wireless Propagation Letters*, vol. 7, pp. 753–756, 2008.

12. C. Yu, W. Hong, and Z. Kuai, "Multiple stopbands ultra wide band antenna," in *2008 International Conference on Microwave and Millimeter Wave Technology*, Nanjing, China, 2008, vol. 4, pp. 1872–1874.

13. M. J. Ammann, "Control of the impedance bandwidth of wideband planar monopole antennas using a beveling technique," *Microwave and Optical Technology Letters*, vol. 30, no. 4, pp. 229–232, 2001.

14. T. Peter and R. Nilavalan, "Study on the performance deterioration of flexible UWB antennas," in *2009 Loughborough Antennas & Propagation Conference*, Loughborough, UK, 2009, pp. 669–672.

15. V. Lakafosis, A. Rida, R. Vyas, L. Yang, S. Nikolaou and M. M. Tentzeris, "Progress towards the first wireless sensor networks consisting of inkjet-printed, paper-based RFID-enabled sensor tags," *Proceedings of the IEEE*, vol. 98, no. 9, pp. 1601–1609, 2010.

16. H. R. Khaleel, H. M. Al-Rizzo and A. I. Abbosh, "Design, fabrication, and testing of flexible antennas," http://dx.doi.org/10.5772/50841.

17. B. M. Sa'ad, S. K. A. Rahim, T. Peter, M. S. B. A. Rani, S. F. Asurodin, D. N. A. Zaidel and C. Krishnan, "Transparent branch-line coupler using micro-metal mesh conductive film," *IEEE Microwave and Wireless Component Letters*, vol. 24, no. 12, pp. 857–859, 2014.

18. B. M. Sa'ad, S. K. A. Rahim and T. Peter, "Transparent 0° phase shifter using micro-metal mesh conductive film," *Electronics Letters*, vol. 51, no. 11, pp. 841–843, 2015.

Chapter 5

Printed Flexible Reconfigurable Wideband Antenna Design for Hybrid Wireless Communication Systems

Mohamed Azab,[a] Bassem Mokhtar,[b] and Ahmed Mansour[b]

[a]*Informatics Research Institute,*
City of Scientific Research and Technological Applications, Alexandria, Egypt
[b]*Department of Electrical Engineering, Alexandria University,*
Alexandria, Egypt

bmokhtar@alexu.edu.eg

5.1 Introduction

There is a huge demand for wireless devices that can provide high data rates over different wireless communication systems (WCSs) with reliable quality of service (QoS) requirements. Such devices should have the ability to perform multi-function and can operate over different WCSs. A reconfigurable antenna is considered an efficient solution to exhibit communication capability in heterogeneous environments. A pixel-reconfigurable antenna offers a low profile, compact size, and reconfigurable shape.

This chapter presents and discusses a set of designs for wideband antennas. The effect of having a single antenna element

Flexible Inkjet-Printed Antennas: Design, Fabrication, and Applications
Edited by Nader Shehata

ISBN 978-981-4968-87-4 (Hardcover), TBA (eBook)
www.jennystanford.com

DOI: 10.1201/9781003774754-5

with different configurations is useful for dynamically adapting over several communication bands in IoT environments and cognitive radio networks. For reaching near-optimum design and operation, optimization algorithms can be used, such as genetic algorithms, for selecting the optimum configuration and operating condition. These antenna designs with related configurations provide different operating frequency bands for RFID, GSM, and WLAN, e.g., 300 MHz, 400 MHz, 900 MHz, 1.2 GHz, 1.6 GHz, 2.4 GHz, 5.2 GHz, 5.8 GHz, and 9.8 GHz with tunable bandwidth. Moreover, the obtained results were used as a set of preliminary simulation studies with an Internet of Things (IoT), cognitive radio network, and WLAN system. These studies show enhancements in the average data throughput and the received signal strengths.

5.2 Printed Reconfigurable Antenna for Hybrid Wireless Communication Systems

Current research in communication networks is to find solutions to dynamically adapt wireless devices with run-time channel variations for optimizing processing time and maximizing data rate over different bands. An effective solution is considered by combining hybrid wireless communications systems (HWCS) to provide efficient devices. This will exhibit a heterogeneous communication environment, which is achieved by incorporating advanced techniques in electronics and wireless communication systems [1].

The design of compact wideband, flexible, and low-cost printed monopole antennas for wireless communication systems that link devices and provide a high-speed connection has taken the interest of a lot of researchers and become the most important part of notebooks in the last decade [2]. Many compact printed antennas are indispensable for different wireless applications that require high-speed transmission of multimedia, image, speech, and data communication, such as ultra-wideband (UWB), wireless local area network (WLAN), and radio frequency identification (RFID) tags, because of the compact size of the antenna, low cost, lightweight, flexible, low profile, and simple fabrication methodology [3–6].

IoT is used to present HWCS, as it is based on integrating the physical and digital components together to create new intelligent products. The idea of IoT is very useful for wireless applications, which have different runtime operating requirement situations. Recently, IoT researchers have focused on developing efficient configurations of wireless sensor networks and antenna designs. Such developments will aid in providing high data rates over reliable HWCS supporting diverse applications with various QoS requirements [5].

Many IoT applications require a combination of 2G/2.5G/3G, Wi-Fi, Bluetooth, NFC, or GPS antennas, which means additional bands and more antennas. It is typically challenging to find enough space to support multiple antennas in one device. Thus, there is a tremendous need for a reconfigurable multi-functional antenna that can adapt to runtime variations, operate over several wireless systems, and have a compact size.

Reconfigurable antennas have gained a lot of attention due to their fitness to redistribute their RF current over the surface, which reflects on their performance. This gives the ability to dynamically serve different applications at the same time. Moreover, combining the concept of reconfigurability with flexible antenna designs will provide an efficient antenna. Such flexible reconfigurable antennas have a low profile, simple inerrability, and a cost-effective fabrication process with reconfiguration capability. It can be used in highly dynamic heterogeneous networking environments, such as IoT and cognitive radio networks [6–8].

The concept of reconfigurability takes different forms and can be applied to different shapes of antennas by dividing them into small parts called pixels. Pixel antennas are basically composed of a grid of electrical small metallic patches, where each pair of adjacent patches is interconnected by a switch. This will enhance the antenna performance by reconfiguring their characteristics based on the activation/deactivation of these switches by (ON/OFF). Common types of reconfigurable antennas use a few number of RF switches, unlike, pixel antennas that use a large number of RF switches [9, 10].

RF switches are the significant elements in all reconfigurable systems. They control changes in antennas to produce different antenna structures with each switching case (ON/OFF). These switching cases reflect changes in the antenna metrics such as frequency, impedance, and radiation. Therefore, the suitable selection of RF switches is important in designing reconfigurable antennas. Moreover, several works [9–15] propose different designs of reconfigurable pixel antennas and some of them use optimization algorithms. Such algorithms are useful to determine the status of each RF switch to synthesize the required frequency band [11].

Thus, a reconfigurable antenna provides a trade-off between design complexity and performance, which mainly depends on increasing reconfiguration capability. A solution is achieved by dynamically changing the architecture of a reconfigurable antenna, which enhances the impedance and radiation characteristics. Such dynamic architectures are mainly based on the size of the pixel, and the interconnection between pixels [15–17]. This motivates the development of designing non-uniform geometries to achieve the required performance with suitable switch numbers [12]. This multi-size reconfigurable antenna architecture allows efficient reconfigurable capabilities for a single antenna. Such a novel reconfigurable single antenna design provides the ability to tune its characteristics, easy impedance matching, and reduces the number of RF switches [12, 13].

Achieving a small size design, easy fed, and low-cost antenna for multiband applications that support different UWB frequencies, which can be applied for IoT, WLAN, WiMAX, digital cellular system (DCS), and UMTS systems is very difficult to achieve [2, 5]. Since common types of UWB antenna designs are microstrip slot antennas with coplanar waveguide fed or microstrip lines have been used to design small, compact designs to increase the bandwidth [18, 19]. Also, the fractal slot and the stub feed line are considered conventional methods for enhancing the bandwidth [20–23].

Regarding printing technique, inkjet is a fast and simple technique and does not require a clean room for rapid and efficient printing electronics. Inkjet printing on a flexible substrate is perfect, as it jets the single ink droplet from a printer's nozzles to a desired position, so there is no waste unlike the etching technique [24]. Using silver nanoparticles inks in the inkjet-printing process ensures higher metal conductivity, as it has been selected to implement circuits and improve the conductivity depending on their higher surface-to-volume ratio with high accuracy and enhanced thermal and electrical conductivity, compared to the bulk-silver ink [25].

5.3 Compact Reconfigurable Multi-Size Pixel Antenna for IoT Environments and Cognitive Radio Networks

This section presents and discusses a novel design of compact flexible reconfigurable multi-size pixel antenna. The proposed antenna provides the capability of adopting its shape in dynamic hybrid wireless communication systems and environments, such as IoT and cognitive radio networks. In the study, in this section, it has been used to serve GSM and WLAN frequency bands (900 MHz, 2.45 GHz, and 5.2 GHz). A simple model of genetic algorithm (GA) is used as an optimization algorithm to provide the capabilities of frequency tuning and bandwidth adjustment based on controlling the status of each RF switch.

This study considers the design of an antenna, which is divided into four quarters, studying the effect of enabling one operating quarter. Moreover, this section compares the proposed multi-size pixels' antenna and same-size pixels' antenna, showing the improvement in the return power loss and power radiation. The presented design provides the ability to assign independent configurations, making the antenna suitable to operate over different frequency bands in cognitive radio networks.

5.4 Flexible Reconfigurable Antenna

Relatively narrow-band antennas with switchable properties are the best solution when the antenna size, frequency reconfigurability, and efficiency are important issues. A single reconfigurable can perform multifunction, improve system performance, give the ability to tune bandwidth, and independently adjust the frequency of operation based on activation/deactivation of some parts of the antenna shape. Moreover, notch antennas are required to obtain the ultimate mobile information sensing.

Reconfigurable antennas consist of a grid of small metallic patches (pixels) and each adjacent patch is connected by an RF switch [14, 15]. RF switches play a significant role in the modification of the current distribution over the antenna surface, which modifies the frequency band, efficiency, and direction of radiation. Their properties reflect on the antenna performance metrics by adding/shifting/removing the operating frequency, tuning bandwidth, and improving directivity and impedance [26–34]. The most common switching components are PIN diodes, FET transistors, and MEMS switches. These components are used to electronically change the frequency response [26, 35–37], radiation patterns [38], gain, or a combination of different radiation parameters of such antennas.

Since the designs of reconfigurable antennas are based on reconfiguring of impedance or radiation, the impedance reconfiguration reflects the bandwidth and resonant frequency tuning. This can be achieved by the electrical modification of antenna dimensions using RF switches [39, 40]. While the radiation reconfiguration controls the polarization and radiation pattern of antennas, common techniques are used to achieve that considering parasitic elements modification [41, 42]. Therefore, the design of a multi-size pixel antenna is an efficient approach to combine two design techniques. In addition, it reduces the requirements of the front-end filter compared to a UWB or multi-band antenna.

The manufacturing process often provides limited options for antenna fabrication for easy integration and fit with different devices. Therefore, there is a growing interest in manufacturing electronics on flexible substrates. Designers consider some parameters to obtain optimum performance and overcome the limitations of the fabrication process. Such parameters are based on a suitable selection of substrate materials, printing techniques, printed inks, and thermal treatments. Inkjet printing is a non-contact printing technology, and it becomes possible to print different patterns on a substrate directly with accurate droplets ejected from a nozzle.

Inkjet-printing technology has a low cost, simple fabrication process, environment-friendly, and less material losses [43]. Flexible electronics can be inkjet-printed on flexible substrates such as polyethylene, terephthalate, polyimide, and paper [44–46]. Inkjet-printing technology has been used in various RF applications such as radio-frequency identification (RFID) tags [47], ultrawideband antennas [48], sensors [49], frequency inductors and capacitors [50], frequency selective surfaces (FSS) [51], transmission lines [52], flexible metamaterial absorbers [53], terahertz split ring resonators [54], and artificial magnetic conductor (AMC) structure [55]. Therefore, this trend provides the development of antennas to be flexible, low-cost, compact size, lightweight, and easy to integrate into devices.

5.5 Internet of Things (IoT) Considerations and Challenges

IoT opens a new era of data communication, device networking, and ubiquitous computing outside the traditional internet realm by virtually presenting every physical thing in the world in a computer version. IoT provides a highly dynamic networking context where there is a massive number of communicating and cooperating mobile devices. Such devices generate, route, and receive a huge volume of data, which are related to various running applications with diverse QoS requirements and communication needs [31, 32].

Numerous applications are applied based on IoT technologies to provide smart solutions across all life domains, such as home automation, autonomous security systems, and smart energy harvesting [33, 34].

One of the main challenges in IoT to suit running various applications is developing efficient communicating devices' antennas with embedded smart context-aware functionalities enabling runtime readjustment and configuration due to: (1) varying applications' needs; (2) available communication bands; and (3) changing operating conditions over time. Therefore, application-oriented reconfigurable wideband antennas are considered an efficient solution that can be used to serve and fit the operation in IoT environments. Such cognitive antennas require the implementation of optimization algorithms, which allow efficient channel access taking into consideration various communication-related parameters, such as operating frequency, required bandwidth, and return power loss. Hence, wireless IoT devices will be able to adopt a single antenna enabling operation in a wide range of communication bands including, for example, GSM (900 MHz) and WLAN (2.4 GHz and 5.2 GHz) with specific bandwidth and radiation patterns.

Furthermore, IoT applications' data characteristics can drive the reconfiguration process of the adopted antenna to meet applications' QoS requirements. As known, various data types and features might exist related to the type of running applications. For instance, video streaming and VoIP applications have different communication bandwidths and delay requirements. Therefore, besides searching in a cognitive way for a suitable communication band, adjusting the antenna configuration pattern considering the applications' data features and QoS requirements enables providing a convenient bandwidth and transmission power supporting efficient secure wireless communications.

Designing reconfigurable antenna modules for IoT devices has some requirements. Most IoT devices are tiny devices with constraints on the available resources. So, some antenna technologies have emerged for IoT, such as chip and PCB antennas [31, 33]. Such antenna technologies have common characteristics like compact size and low cost. However, there are some features that impede the antenna operation with a wide range of frequencies, such as the fixed antenna design area and the used substrate material. Accordingly, we will discuss in upcoming sections the proposed antenna design and related analysis, which shows how such design suits the usage of antennas in an IoT environment.

The next sections discuss the proposed antenna design and the methodology for developing a near-optimum design adopting some optimization algorithms. Such algorithms are used to reconfigure the antenna design to enable antenna operation with various communication bands and diverse types of applications with different QoS needs. A methodology is considered for refining the interaction between the antenna hardware and controlling software, issuing a software-defined reconfigurable antenna.

5.5.1 Near-Optimum Antenna Design Methodology

Optimization is a process that is used to adjust input parameters to find the desired output quantities achieving high performance. Several optimization schemes have been implemented for antenna designs such as genetic algorithm (GA), practical swarm optimization (PSO), and artificial neural network (ANN) [27]. Those powerful algorithms are searching to find the status of each RF switch to adjust the antenna to a specified value/range of frequencies. A genetic algorithm has been used due to its good performance. It is a powerful optimization algorithm regarding convergence speed and robustness for achieving global minima when applied to electromagnetic problems.

GA is based on converting interested input parameters of the given fitness function into chromosome geneses. After selecting the preferred chromosomes according to the required fitness (output) each pair of chromosomes (parents) is cross-over. Then the mutation generates a new offspring that carries mixed properties of both parents. This fitness calculation is repeated till the maximum number of generations is reached, or the chromosome with maximum fitness is achieved (optimum solution).

$$F = \max_{x_i \in (ON,\ OFF),\ i = 1,2...} [S_{11}\,(f_i) - 10 \log (1 - |S_{11}|^2)] \geq -20 \text{ dB} \quad (5.1)$$

where f_i is the value of operating frequency. In the case of 2.45 and 5.1 GHz, $S_{11}(f_i)$ is the return loss amplitude at operating frequency f_i, and x refers to the status of each RF switch.

The antenna designs are implemented on MATLAB and the results are calculated via an electromagnetic simulations software tool such as Ansoft high-frequency simulation software (HFSS). This technique is called MATLAB-HFSS scripting and it is useful for calculating the fitness of each solution of the optimization algorithm GA. The results are evaluated by exporting the data from MATLAB to simulate it at HFSS, then sent back to MATLAB to verify and assign optimal design parameters.

Pseudo-code for the design process

1. Write MATLAB Scripting Code for the proposed antenna design.
2. Randomly initialize a population of individuals within the constraint range.
3. Assign the initial configuration of each RF switch.
4. Evaluate the fitness of each initial individual according to the fitness function via HFSS after sending the results back to MATLAB.

While *(current generation < number of generation)*

5. Sort the populations in ascending order based on their fitness values.
6. Choose the top populations and apply the crossover and mutation process.
7. Calculate the fitness values of the new generation populations and select the best population.
8. Assign a new configuration for each RF switch.
9. Evaluate the fitness of each new individual via MATLAB-HFSS interface environment.
10. Repeat steps from 6 to 10 till the condition is terminated.

5.5.2 Antenna Geometry

The proposed antennas are simple rectangular microstrip patch-printed antennas. They are designed to operate at a center frequency of 950 MHz, 2.45 GHz, and 5.2 GHz, respectively. Both designs have an overall size of 80 × 95 mm^2 and they are divided into small parts (pixels), as shown in Fig. 5.1. The first design presents the same-size pixel antenna and the patch antenna is divided into 3 × 3 pixels, each one has a size of $L_{\text{strip}} \times W_{\text{strip}}$ = 15 mm × 15 mm, as shown in Fig. 5.1b.

The second design presents the novel multi-size pixels antenna, which depends on changing the size of pixels of both the middle column and row. Figure 5.1a shows the proposed multi-size pixel antenna divided patch antenna into 3 × 3 main pixels with different pixels' sizes to provide more ability to control bandwidth. Both antennas have a ground plane on one side of the dielectric substrate and the pixeled shape is printed on the top layer of the substrate with a thickness of 1.2 mm. The design of the same-size pixel antenna needs 12 RF switches, while the proposed multi-size pixel antenna design needs 44 RF switches. The second design provides the ability to accurately tune bandwidth compared to the first design.

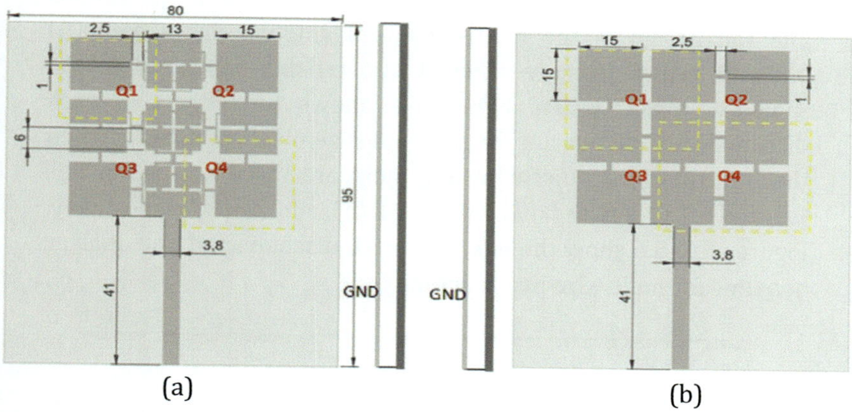

Figure 5.1 Dimension pixel antenna. The full dimension of (a) multi-size pixel antenna and (b) same-size pixel antenna.

5.5.3 Antenna Analysis and Simulation Results

This section presents a simulation that results from the HFSS software tool for different switching scenarios. In simulation, the switches are considered ideal switches and are replaced with metal strips with $L_{strip} \times W_{strip}$ = 0.5 mm × 0.5 mm. Both antennas are simulated assuming glossy paper as a substrate with a relative permittivity of 3.2, and loss tangent of 0.08 [22]. The genetic algorithm (GA) is used to synthesize antenna frequency and the bandwidth with each scenario. Two techniques are presented to compare the effects of both antenna designs. The first technique focuses on the effect of the quarter, as it divides the antenn into four quarters, then disconnects one quarter and keeps the other three quarters connected together. This technique is applied on both antenna designs and the results are similar for both antennas. Figure 5.2 shows the S_{11} of both designs when all quarters are connected. It can be noticed that there are notches at 2.4, 3.3, 5.2, and 8.2 GHz for both antennas, while the same-size pixel antenna has more notch at 1.2 GHz. Also, in the case of a multi-size pixel antenna, the bandwidth span is wider and the amplitude of return loss indicates efficient matching. Figure 5.3 shows the 3D radiation pattern of both antenna designs and the design of the multi-size pixel antenna has a gain (G) of 5 dB, which is more efficient and reliable compared to the same-size pixel design. Figure 5.4 shows the effects of each disconnected one quarter for the same-size pixel antenna, as when the first quarter is disconnected, there are three notches at 0.9, 2.5, and 5.1 GHz with bandwidth 0.04, 0.3, and 2 GHz, respectively. Figures 5.3–5.6 show the effects of each disconnected one quarter at a time for multi-size pixels antenna.

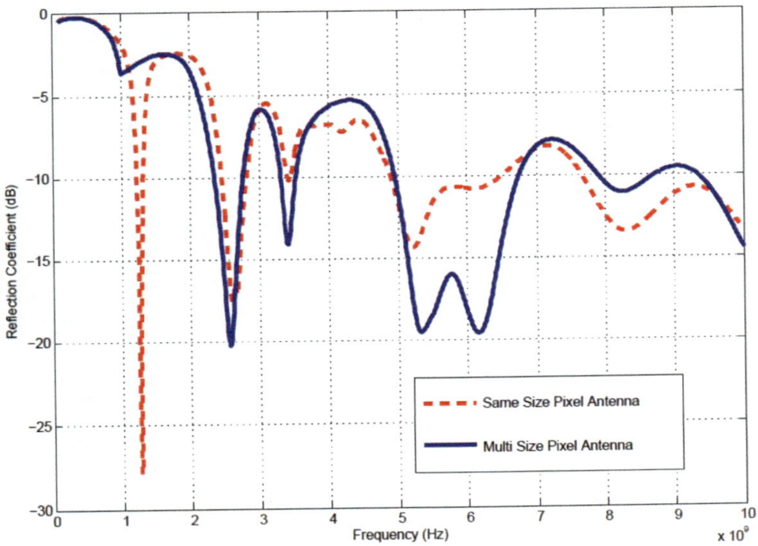

Figure 5.2 Simulated S_{11} of SS pixel antenna and MS antenna.

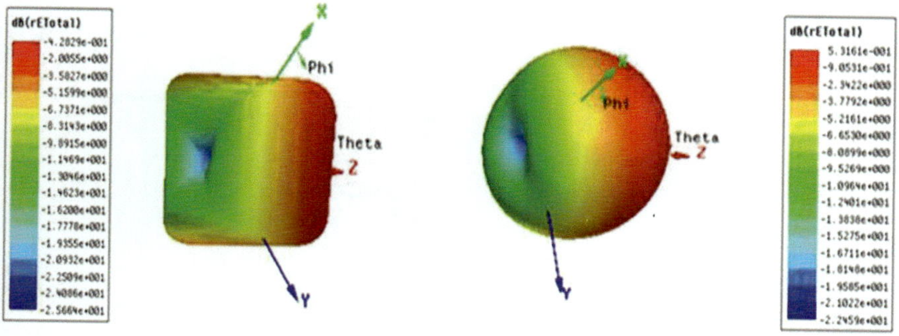

Figure 5.3 3D radiation pattern of the proposed pixel antennas (a) same-size pixel antenna; (b) multi-size pixel antenna.

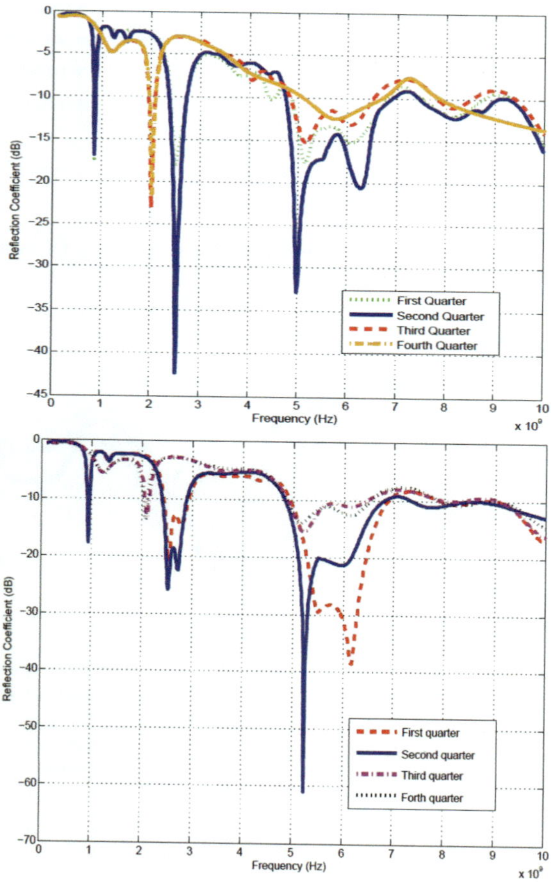

Figure 5.4 Simulated S_{11} of same-size pixel antenna when one quarter is disconnected.

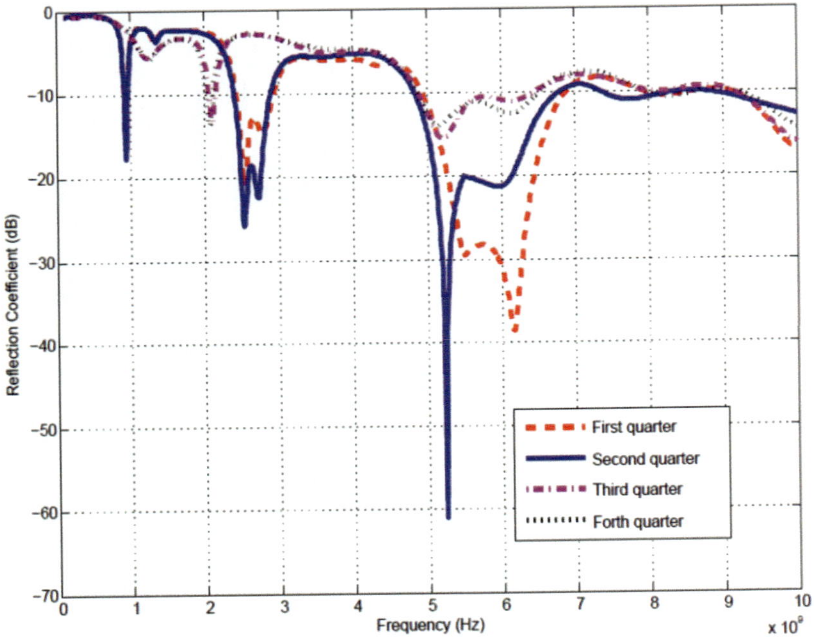

Figure 5.5 Simulated S_{11} of MSPRA when one quarter is disconnected.

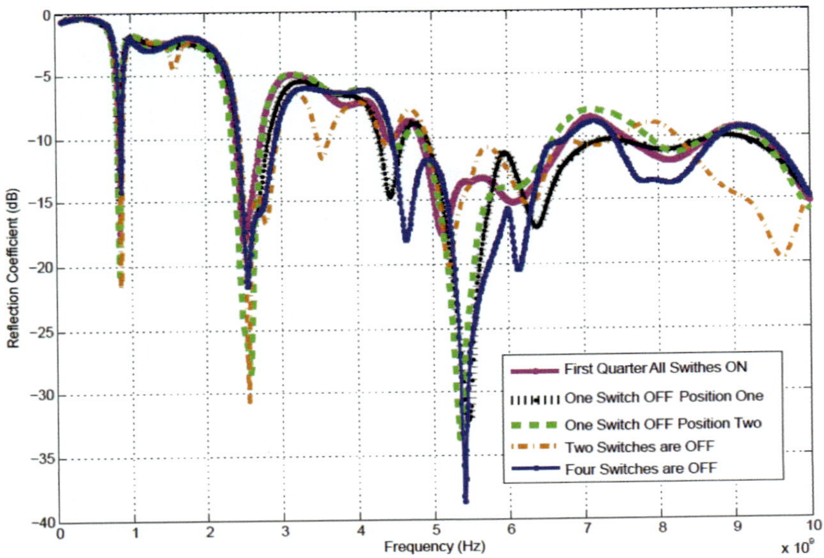

Figure 5.6 Simulated S_{11} of MSPRA when the Q1 is disconnected.

When the second quarter is disconnected, the operating frequency is massively affected and its operating frequencies are the same as the first quarter, but the bandwidth of each frequency is 0.05, 0.4, and 2.2 GHz. Figure 5.5 shows the effects of each disconnected quarter for a multi-size pixel antenna, as when the first quarter is disconnected, there is a notch at 2 GHz. When the second quarter is disconnected, there are several notches at 0.98, 2.5, 5, and 6.2 GHz, and when the third or fourth quarter is disconnected, there are narrow bandwidth notches at around 2.4 GHz and 5.4 GHz. The second technique shows the advantage of multi-size pixel antenna design, which is useful to control the tuning range. This technique investigates the effect of turning off each RF switch of the deactivated quarter. This will enhance the impedance matching, tune bandwidth, and reduce the biasing voltage of used RF switches. The following figures summarize the effect of tuning bandwidth with the output from the GA optimization algorithm based on five different scenarios for each disconnected quarter. The first scenario considers all switches of the disconnected quarter when they are ON. The second scenario shows the effect when there is only one switch is OFF. The third scenario is the same as the second scenario, but it considers a different switch to switch it OFF. The fourth scenario considers two switches with two new different positions when they are OFF. The final scenario shows the effect when all the mentioned switches above are OFF at the same time. Figure 5.6 shows that a fixed notch almost at 0.9 GHz appeared in all scenarios with the same return loss (RL), while there is a slight shift in the notch at 2.5 GHz giving different RL and bandwidth spans. Figure 5.7 shows a notch at 0.9 GHz, the same as shown in Fig. 5.8, and two shifted notches appeare between 2.4 and 2.7 GHz, and 5 and 5.6 GHz, respectively. Figure 5.8 shows different operating frequencies, as there are two main notches at 3.8 and 5 GHz, in the case of the second scenario. Also, there are two new notches at 2 and 5 GHz in the case of the third scenario. Figure 5.9 shows that there are also different operating frequencies at 5.2, 2, and 8.5 GHz, which are obtained with different RF configuration scenarios. All results of different operating frequencies and bandwidth ranges for each quarter are summarized in Table 5.1.

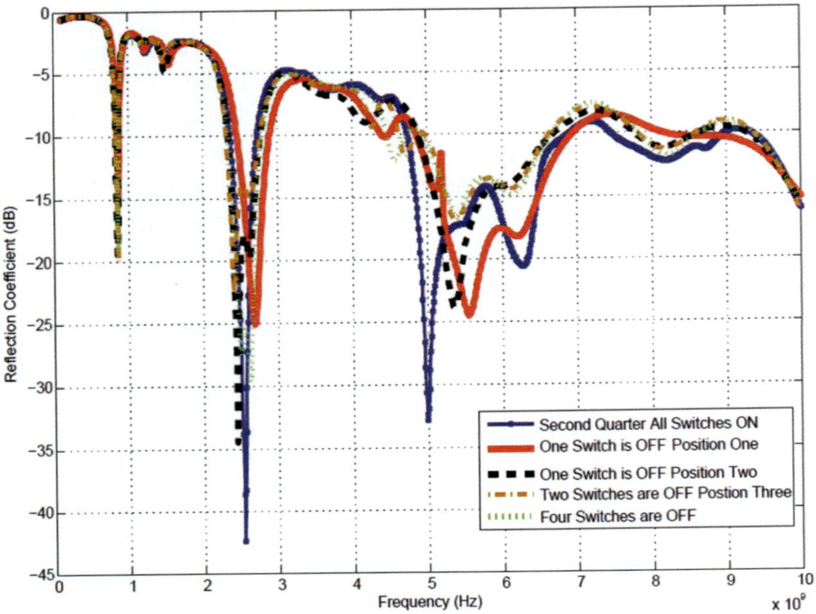

Figure 5.7 Simulated S_{11} of MSPRA when the Q2 is disconnected.

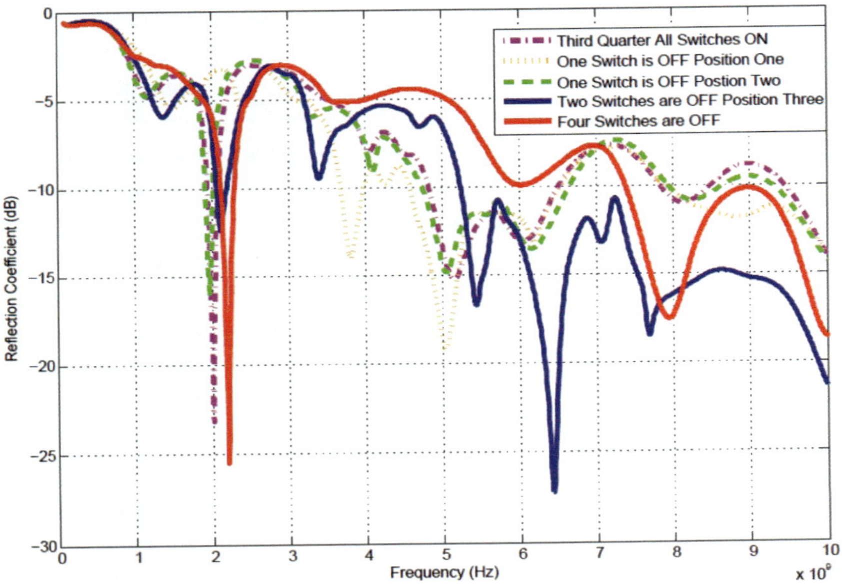

Figure 5.8 Simulated S_{11} of MSPRA when the Q3 is disconnected.

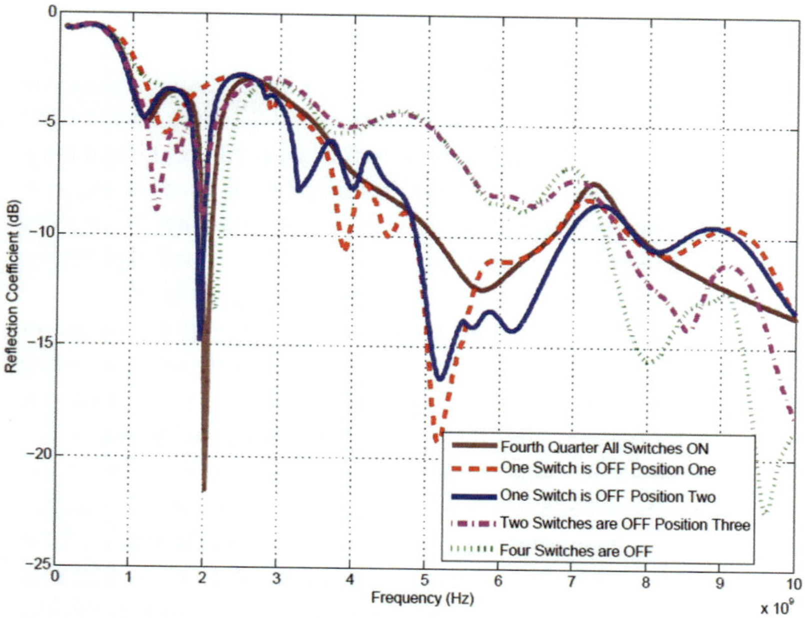

Figure 5.9 Simulated S_{11} of MSPRA when the Q4 is disconnected.

Table 5.1 Operating frequency and bandwidth range of each quarter at different switch configurations

Switches Configuration	First Quarter f_0 (GHz)	First Quarter BW (GHz)	Second Quarter f_0 (GHz)	Second Quarter BW (GHz)	Third Quarter f_0 (GHz)	Third Quarter BW (GHz)	Fourth Quarter f_0 (GHz)	Fourth Quarter BW (GHz)
All Quarter Switches ON	0.86	0.84 - 0.88	0.85	0.83 –0.88	2.02	1.9 – 2.08	2.03	1.96 – 2.09
	2.5	2.4 – 2.7	2.5	2.4 – 2.7	5.2	4.8 – 6.5	5.8	5.1 – 6.7
	5.1	4.8 – 6.8	5	4.7 – 6.9				
One Switch OFF – Position1	0.85	0.83 – 0.87	0.86	0.84 -0.88	3.8	3.7 – 4	5.2	4.8 – 6.7
	2.55	2.4 – 2.8	2.7	2.5 – 2.9				
	4.44	4.3 – 4.6	5.55	4.9 – 6.9	5	4.6 – 6.6		
	5.5	4.9 – 7.4						
One Switch OFF – Position2	0.84	0.82 – 0.86	0.84	0.82 – 0.87	1.97	1.92 – 2.02	5.2	4.9 – 6.9
	2.57	2.3 – 2.7	2.44	2.3 – 2.8	5.1	4.7 – 6.7	1.94	1.9 – 1.99
	5.4	4.9 – 6.6	5.4	5 – 6.6				
Two Switches OFF	0.86	0.84 – 0.89	0.85	0.83 – 0.87	2.12	2.07 – 2.2	8.6	7.6 –10
	2.55	2.44 – 2.94	2.4	2.3 – 2.75				
	5.2	4.9 – 6.8	5.4	4.9 – 6.7	6.5	5.2 –10		
	9.7	8.2 –10						
Four Switches OFF	0.87	0.85 – 0.89	0.85	0.83 – 0.87	2.21	2.13 – 2.3	9.6	7.4 –10
	2.55	2.4 – 2.9	2.6	2.34 – 2.78				
	5.4	4.5 – 6.8	5.4	4.9 – 6.7	8	7.3 –10	2.14	2 – 2.22
	8.1	7.4 – 8.8						

5.5.4 Network Simulation Study

The proposed MSPR antenna is used in a simulation scenario for communicating between network nodes, and it is assumed to offer data transmission using various radios and different frequency bands. A cognitive radio network simulator is used to support performance evaluations for the proposed antenna. It enables dynamic spectrum resource allocation, power control algorithms, and running adaptive cognitive radio networking protocols.

This software is a network-based simulator, and it is based on the network simulator 2 (ns-2). The simulated network assumes six communicating nodes. There are three source nodes that run file transfer services, thus there are three distinct data flows.

Simulation parameters are: topology size = 50 m × 50 m, number of nodes = 6, maximum queue size = 50 packets, number of sources = 3, number of destinations = 3, traffic type is FTP, packet size = 512 bytes, MAC layer protocol is IEEE 802.11b, channel bandwidth = 2 MBytes, simulation time = 50 sec, routing protocol is DSR, and pause time = 10 s.

Each node, including the transmitters and receivers, adopts the reconfigurable antenna, which enables data transmission and reception over various bands. Each node is aware of the available communication bands and related channels to satisfy the QoS requirements of running applications. Therefore, each node searches for a suitable communication band where the primary user does not work. Then, the node reconfigures different configuration patterns of its antenna by firing some switches to provide a suitable bandwidth and return power loss to maximize data throughput. The target is measuring the average data throughput as a performance metric to evaluate the usage of the proposed antenna. The following two cases are considered:

- The available free band for each node is one band. Thus, each node is assumed to provide the ability to transmit and receive using one radio channel.
- Three available communication bands are free such that nodes can switch between those bands to maximize the data throughput.

Figure 5.10 shows the average data throughput at having one radio or a communication band per communicating node. The obtained average throughput is 51,349 bytes. Moreover, the average data throughput at having three radio or communication bands per communicating node is based on the designed antenna at the physical layer. The obtained average throughput is 52,133 bytes. The results show implementing a wideband reconfigurable antenna, enables communication over various bands that will offer higher opportunities for each node to communicate and provide support for meeting the QoS requirements of running applications.

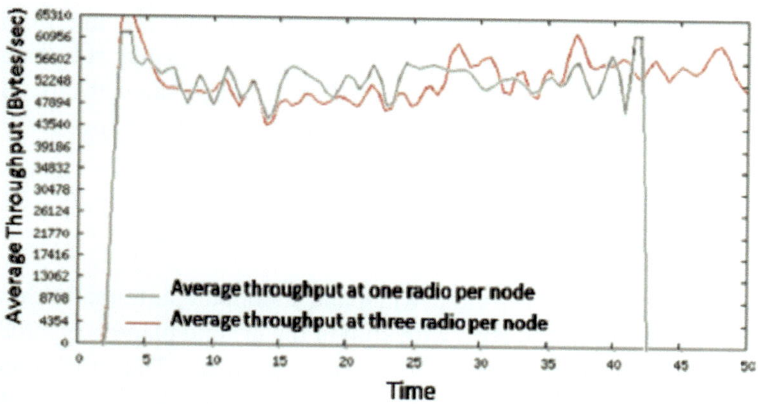

Figure 5.10 Average data throughput per node.

5.6 Wideband Printed Monopole Paper-Based Antenna for Ultra-Wideband Applications

In this section, we introduce a normal desktop inkjet printer with silver nanoparticle ink to enhance the bandwidth of a simple microstrip-fed-folded strip monopole antenna with a protruding stub in the ground plane. This antenna was introduced before in [48] printed on a rigged substrate with frequency selective notch at 2.4 GHz and 5.8 GHz. However, our modification of printing an inkjet flexible antenna has achieved a wideband performance to simultaneously cover the most common commercial frequencies that are used for WLAN and WiMAX bands. The remaining part of this section also provides detailed antenna dimensions and comparisons between simulated and measured results based on inkjet printing onto different substrates (glossy paper and PET film).

5.6.1 Fabrication Process

A lot of antenna designs have been fabricated on flexible substrates such as papers, textiles, and PET films. The effect of flexibility is discussed showing the enhancement in performance [51–55]. We focused on photo paper (glossy) and PET films as they are suitable for inkjet electronics printing. Also, they were exposed to some chemical coating treatments to prevent the ink from absorption and adhere it to paper [52]. The flexible substrate has a thickness equal to 0.25 and 0.26 mm, a loss tangent of 0.05 and 0.008, and a relative permittivity of 4.01 and 3.2, respectively, for glossy paper and PET film.

5.6.2 Antenna Design

A flexible and inkjet-printed wideband monopole antenna with a microstrip-fed-folded strip with stub in the ground plane is printed on the glossy paper and PET film as substrates. The fabricated prototype antenna with full dimension is shown in Fig. 5.11. A 50 Ω microstrip line is used for the excitation, and it has an overall area of 35 × 37.5 mm^2.

The antenna is connected with an SMA connector using silver epoxy with a ratio of 1:1 and is dried at room temperature. The folded strip width and protruding stub width equal the width of the microstrip line. This design was proposed in [48] on an RF-4 substrate to support only two resonances at 2.4 GHz and 5.8 GHz for WLAN and RFID applications, but in this study, a photo paper (glossy) and PET films are used as a substrate with silver nanoparticles ink to enhance the performance and support multiband frequency.

Varying lengths of resonance paths (L1, L2) for the folded strip and protruding stub will highly affect antenna resonance frequencies by a little bit of shifting or rejection or changing on available bandwidth in the case of printing on the ragged substrate [48]. But suitable lengths with flexible inkjet printing will support different existing wireless network frequencies easily independent on (L1, L2), which can be used with routers by replacing normal monopole antenna with our inkjet flexible antenna.

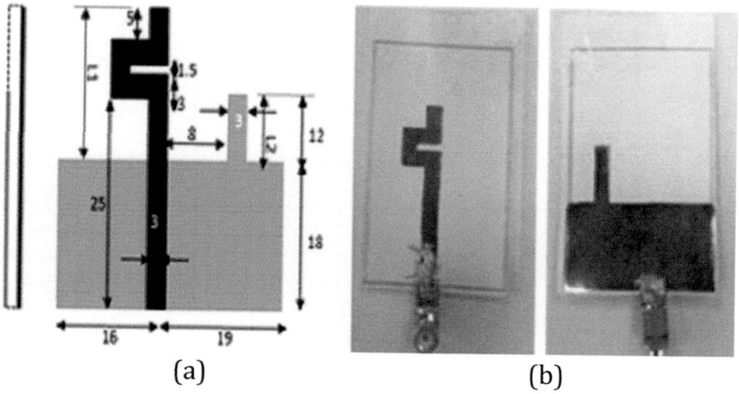

Figure 5.11 The proposed WLAN printed antenna: (a) Geometry of front and back side antenna; (b) Fabricated prototype of antenna.

5.6.3 Analysis and Results

This antenna design is simulated using Ansoft high-frequency simulation software (HFSS). This section presents the simulated and experimental work of the proposed antenna from a network analyzer and compares it by simulation analysis using Ansoft HFSS for both antennas printed on glossy and PET film, resulting in the return loss (S_{11}), which can be summarized in Fig. 5.12. It can be observed that the higher performance of S_{11} can be attained by using PET film and glossy paper as substrates than FR-4 for the simulated cases. Moreover, the effects of multilayer printing and thermal treatments cause a wideband frequency, as shown in Fig. 5.13. This multilayer wideband antenna is suitable to operate at several bands of different wireless communication systems. For example, WLAN (5.15–5.825 GHz), (2.4–2.48 GHz), WiMAX (5.25–5.85 GHz), DCS (1.71–1.88 GHz), PCS (1.75–1.87 GHz), and UMTS (1.92–2.17 GHz), according to IEEE 802.11 standard.

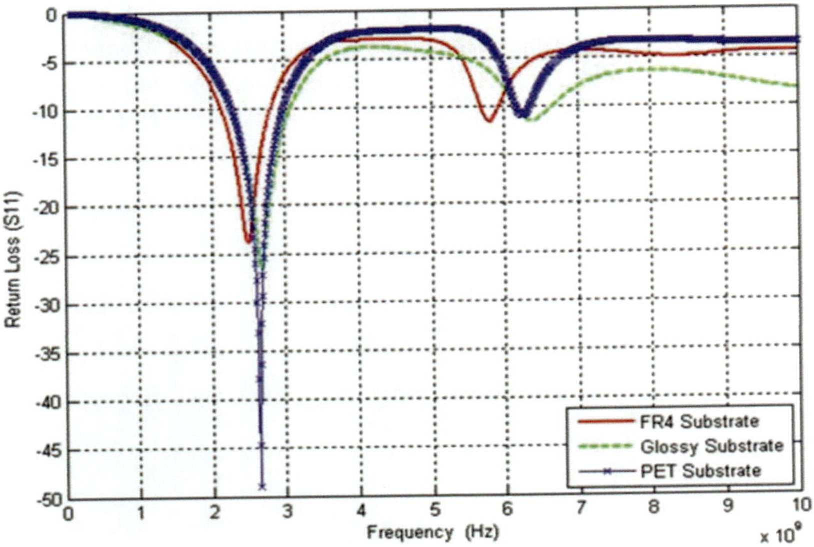

Figure 5.12 Simulated return loss S_{11} of the proposed antenna on different substrates.

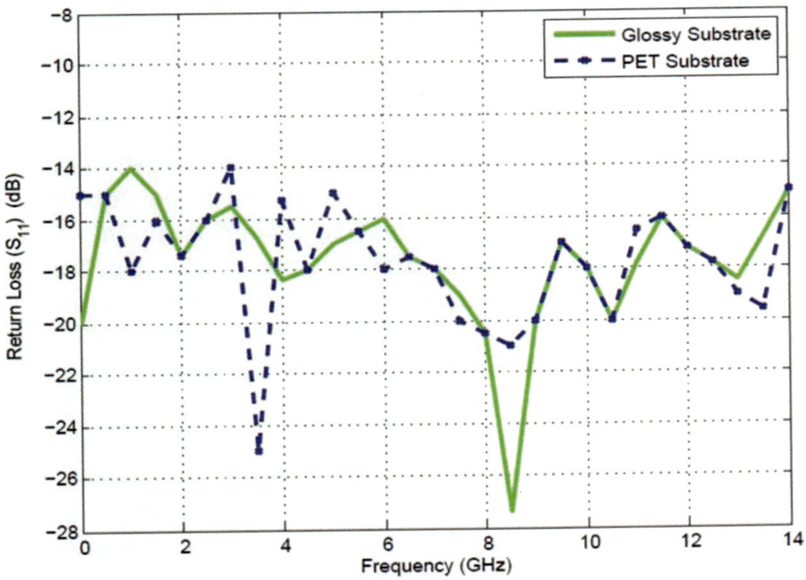

Figure 5.13 Measured return loss S_{11} of the proposed antenna on different substrates.

Figure 5.14 shows the 3D model of the radiation pattern of the printed antennas. It can be noticed that the gain is enhanced in the case of antenna design printed onto glossy paper and PET film compared to printed antenna onto FR-4 substrate.

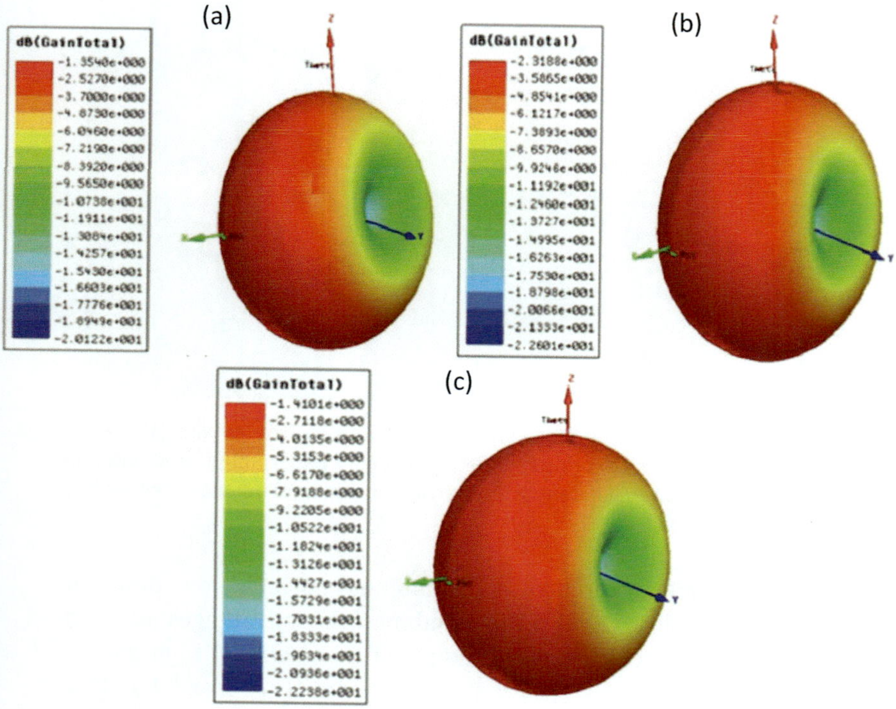

Figure 5.14 Simulated 3D model of the total gain in (dB) of the proposed antenna on different substrates: (a) 3D radiation of FR-4 substrate; (b) 3D radiation of glossy substrate; (c) 3D radiation of PET film substrate.

5.6.4 Practical Application

Based on the previous analysis, we replace the normal monopole antenna of the router with the proposed paper-based inkjet-printed antenna. We found the transmission and receiving process of data within 4 m away from the router has a reception rate of 100%. During this test, the transmission and reception process of data has a reception rate of 100% within 4 m away from the router. Figure 5.15 shows different antenna configurations with the router. This measurement is done using a free Android program called "Wi-Fi signal strength", and any wireless router such as a TP-link router. This test operates on a frequency equal to 2.45 GHz.

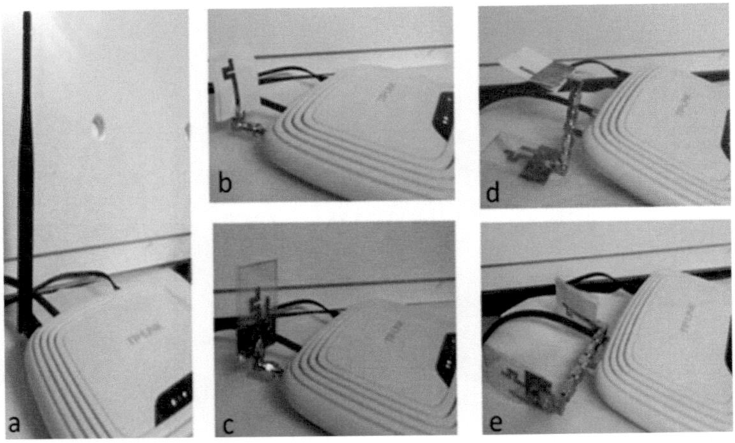

Figure 5.15 Different antenna configurations with the router with (a) normal monopole antenna; (b) printed antenna on glossy paper; (c) printed antenna on PET film; (d) double antennas in a vertical position; (e) double antennas in a horizontal position.

The configuration of two antennas printed on different substrates (glossy and PET) combines the advantages of each type and enhances the performance of the router. A high gain is obtained by using a glossy paper antenna (increased by 30% compared to FR-4), and minimum loss is achieved due to the PET film antenna.

Figure 5.16 shows changes in signal strength in (dBm) for different antenna configurations with distance in meters. It could be noticed that in the case of a small distance of less than 30 cm, the signal strength of a flexible inkjet-printed antenna on PET film is better than a normal monopole router antenna. Moreover, with the usage of two antennas (horizontal or vertical position), the performance can be enhanced and become close to a normal monopole router antenna in the case of large distances. Furthermore, this performance of the antenna can be enhanced more by heating treatment and multilayer printing to improve the conductivity depending on their higher surface-to-volume ratio and enhance thermal and electrical conductivity [50].

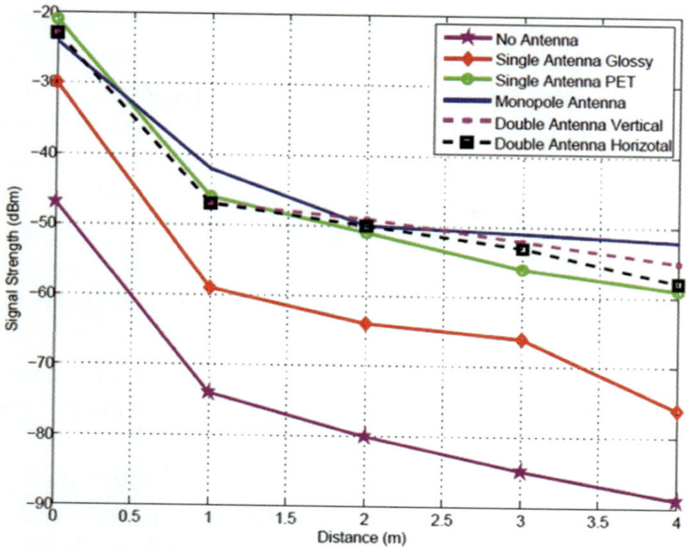

Figure 5.16 Signal strength of different antenna configurations with distance.

5.7 Conclusion

This chapter presented and discussed novel designs of reconfigurable antennas for hybrid wireless communication systems. These designs offer the ability to provide different operating frequencies and control their bandwidth by firing suitable antenna configurations. Moreover, the effects of tuning bandwidth based on optimization algorithms to determine the status of each configured RF switch. These effects are very beneficial to opportunistic networking where there are no static end-to-end paths between sources and destinations due to mobility and limitations in available transmission bands. This feature will allow mobile routing nodes to receive data and adapt to various transmission bands.

According to the obtained results, we notice that multi-size pixel reconfigurable antenna can be used to operate at a specific band or a set of communication bands according to interesting applications. It also provides the ability to tune bandwidth based on the type of transmitted/received data (text/voice/video) by configuring certain RF switches. This is useful for WLAN applications, as the routers will only process and forward mobile data and voice with limited data rates. Moreover, other data transmitted from other bands will not be considered, which can be considered as an advantage for the security system. Besides, this limits the data analysis scale and saves more resources and consuming power. This approach can be used for smart switchable antennas that can be assigned to operate at the required wireless communication system bandwidths, supporting various data types and enhancing the QoS of the applications. In addition to the ability to use in sensing in cognitive radio systems, by selecting the most suitable and free band for communication, with minimum noise or interference, to transmit a specific type of data.

Simulations analysis and experimental results show the use of an inkjet printing antenna on a flexible substrate and they will improve the performance of the same antenna design and enhance its bandwidths, which are wide enough to support common mobile standards such as PCS, DCS, UMTS, GSM1800 and 1900, WiMAX and WLAN/Bluetooth, or in energy harvesting using ambient electromagnetic energy in the form of low frequency. This means the device has a flexible antenna that can be used in different applications in several countries, without any additional cost. Also, this work is based on one of the promising techniques that used silver nanoparticles in fabricating nanoantenna with a feasible and commercial method for printing.

References

1. M. R. Gouvêa, E. C. Belvedere, J. J. Oliveira, P. E. Mascigrande, A. P. Costa, and R. Brunhetoro, "Development of standards for underground hybrid networks (in Portuguese)," *Aneel*, 2005.
2. C. Wang, Z.-H. Yan, P. Xu, J.-B. Jiang, and B. Li, "Trident-shaped dual-band CPW-fed monopole antenna for PCS/WLAN applications," *IEEE Electronics Letter*, vol. 47, no. 4, pp. 231–232, 2011.
3. T.-L. Zhang, Z.-H. Yan, L. Chen, and Y. Song, "A compact dual-band CPW-fed planar monopole antenna for WLAN applications," *Journal of Electromagnetic Waves and Applications*, vol. 22, nos. 14–15, pp. 2097–2104, 2008.

4. J.-L. Jaw, F.-S. Chen, and D.-F. Chen, "Compact dualband CPW-fed slotted patch antenna for 2.4/5 GHz WLAN operation," *Journal of Electromagnetic Waves and Applications*, vol. 23, nos. 14–15, pp. 1947–1955, 2009.
5. K. Zhao, S. Zhang, and S. L. He, "Closely-located MIMO antennas of tri-band for WLAN mobile terminal applications," *Journal of Electromagnetic Waves and Applications*, vol. 24, nos. 2–3, pp. 363–371, 2010.
6. W. S. Chen and Y.-H. Yu, "Dual-band printed dipole antenna with parasitic element for WIMAX applications," *Electronics Letters*, vol. 44, no. 23, pp. 1338–1339, 2008.
7. J. Gubbi et al., "Internet of Things (IoT): A vision, architectural elements, and future directions," *Future Generation Computer Systems*, vol. 29, no. 7, pp. 1645–1660, 2013.
8. D. Rodrigo, B. A. Cetiner, and L. Jofre, "Frequency, radiation pattern and polarization reconfigurable antenna using a parasitic pixel layer," *IEEE Transactions on Antennas and Propagation*, vol. 62, no. 6, pp. 3422–3427, 2014.
9. A. Grau, J. Romeu, L. Jofre, and F. De Flaviis, "A software defined MEMS-reconfigurable pixel-antenna for narrowband MIMO systems," in *2008 NASA/ESA Conference on Adaptive Hardware and Systems*, Noordwijk, Netherlands, pp. 141–146, 2008.
10. L. N. Pringle, P. H. Harms, S. P. Blalock, G. N. Kiesel, E. J. Kuster, P. G. Friederich, R. J. Prado, J. M. Morris, and G. S. Smith, "A reconfigurable aperture antenna based on switched links between electrically small metallic patches," in *IEEE Transactions on Antennas and Propagation*, vol. 52, no. 6, pp. 1434–1445, 2004.
11. D. Rodrigo et al., "Genetic reconfigurability of a multi-size pixelled antenna," in 2010 *IEEE International Conference on Wireless Information Technology and Systems*, Honolulu, HI, USA, pp. 1–4, 2010.
12. D. Rodrigo et al., "Small pixelled antenna with MEMS-reconfigurable radiation pattern," 2010 *IEEE Antennas and Propagation Society International Symposium*, Toronto, ON, Canada, pp. 1–4, 2010.
13. D. Rodrigo et al. "MEMS-reconfigurable antenna based on a multi-size pixelled geometry," in *Proceedings of the Fourth European Conference on Antennas and Propagation*, Barcelona, Spain, pp. 1–4, 2010.

14. A. Grau, J. Romeu, L. Jofre, and F. De Flaviis, "A software defined MEMS reconfigurable PIXEL antenna for narrowband MIMO systems," in *2008 NASA/ESA Conference on Adaptive Hardware and Systems*, Noordwijk, Netherlands, pp. 141–146, 2008.
15. Chen M, Hu H, Lei S, Sun K, Tian J, Chen B. A shared aperture antenna with reconfigurable pattern by common and differential modes. *Microwave and Optical Technology Letters*, vol. 66, p. e33888, 2024.
16. L. N. Pringle, P. H. Harms, S. P. Blalock, G. N. Kiesel, E. J. Kuster, P. G. Friederich, R. J. Prado, J. M. Morris, and G. S. Smith, "A reconfigurable aperture antenna based on switched links between electrically small metallic patches," *IEEE Transactions on Antennas and Propagation*, vol. 52, pp. 1434–1445, 2004.
17. J. C. Maloney, M. P. Kesler, L. M. Lust, L. N. Pringle, T. L. Fountain, P. H. Harms, and G. S. Smith, "Switched fragmented aperture antennas," *IEEE Antennas and Propagation Society International Symposium. Transmitting Waves of Progress to the Next Millennium. 2000 Digest. Held in conjunction with: USNC/URSI National Radio Science Meeting* (C, Salt Lake City, UT, USA), vol. 1, pp. 310–313, 2000.
18. K. Moradi and S. Nikmehr, "A dual-band dual-polarized microstrip array antenna for base stations," *Progress in Electromagnetics Research*, vol. 123, pp. 527–541, 2012.
19. S. Mohammad, H. R. Hassani, and A. Foudazi, "A dual-band WLAN/UWB printed wide slot antenna for MIMO/diversity applications," *Microwave and Optical Technology Letters*, vol. 55, no. 3, pp. 461–465, 2013.
20. L. Wen-jun and ZHU Hong-bo, "Frequency notched wide slot antenna for UWB/2.4 GHz WLAN applications," *The Journal of China Universities of Posts and Telecommunications*, vol. 14, no. 4, pp. 122–125, 2007.
21. H. Eskandari and M. N. Azarmanesh, "Bandwidth enhancement of a printed wide-slot antenna with small slots," *AEU – International Journal of Electronics and Communications*, vol. 63, no. 10, pp. 896–900, 2009.
22. K. Ahmed, K.-F. Lee, C. E. Smith, and D. Kajfez, "Microstrip line and CPW FED ultra wideband slot antennas with U-shaped tuning stub and reflector," *Progress in Electromagnetics Research*, vol. 56, pp. 163–182, 2006.

23. F. B. Zarrabi et al., "Triple band microstrip slot antenna for WIMAX/ WLAN applications with SRR shape ring," in *Proceedings of the 2014 Fourth International Conference on Advanced Computing & Communication Technologies*, pp. 368–371, 2014.
24. J. R. Panda and R. S. Kshetrimayum, "A printed 2.4 GHz/5.8 GHz dual-band monopole antenna with a protruding stub in the ground plane for WLAN and RFID applications," *Progress in Electromagnetics Research*, vol. 117, pp. 425–434, 2011.
25. L. Yang, L. Rida, R. Vyas, S. Bhattacharya, and M. M. Tentzeris, "RFID tag and RF structures on paper substrate using inkjet-printing technology," *IEEE Transaction on Microwave Theory and Techniques*, vol. 55, no. 12, part 2, pp. 2894–2901, 2007.
26. A. M. Mansour, N. Shehata, B. M. Hamza, and M. R. M. Rizk, "Efficient design of flexible and low cost paper-based inkjet-printed antenna," *International Journal of Antennas and Propagation*, vol. 2015, Article ID 845042, 2015.
27. P. J. Rainville and F. J. Harackiewicz, "Magnetic tuning of a microstrip patch antenna fabricated on a ferrite film," *IEEE Microwave and Guided Wave Letters*, vol. 2, no. 12, pp. 483–485, 1992.
28. Y. Rahmat-Samii and E. Michielssen (eds.), *"Electromagnetic Optimization by Genetic Algorithms,"* John Wiley & Sons, New York, 1999.
29. E. Fleisch, "What is the Internet of Things? An economic perspective," *Economics, Management and Financial Markets,* vol. 5.2, p. 125, 2010.
30. O. Vermesan and P. Friess (eds.), *"Internet of Things-From Research and Innovation to Market Deployment,"* River Publishers, 2014.
31. ITU-T, Recommendation G.107, "G.107: The E-model: A computational model for use in transmission planning," Recommendation ITU-T, 2005.
32. J. Gubbi et al., "Internet of Things (IoT): A vision, architectural elements, and future directions," *Future Generation Computer Systems*, vol. 29.7, pp. 1645–1660, 2013.
33. R. Payne and B. Macdonald, "Ambient technology—now you see it, now you don't," *BT Technology Journal*, vol. 22.3, pp. 119–129, 2004.
34. L. Yan and Y. Zhang, "The Internet of Things: From RFID to the Next-Generation Pervasive Networked Systems," Routledge, 2008.
35. F. Wortmann and K. Flüchter, "Internet of things," *Business & Information Systems Engineering*, vol. 57.3, pp. 221–224, 2015.

36. D. Peroulis, K. Sarabandi, and L. P. B. Katehi, "Design of reconfigurable antennas," *IEEE Transactions on Antennas and Propagation*, vol. 53, no. 2, pp. 645–654, 2005.
37. J. T. Aberle, O. Sung-Hoon, D. T. Auckland, and S. D. Rogers, "Reconfigurable antennas for wireless devices," *IEEE Antennas and Propagation Magazine*, vol. 45, no. 6, pp. 148–154, 2003.
38. B. A. Cetiner, H. Jafarkhani, Q. Jiang-Yuan, Y. H. Jae, A. Grau, and F. De Flaviis, "Multifunctional reconfigurable MEMS integrated antennas for adaptive MIMO systems," *IEEE Communication Magazine*, vol. 42, no. 12, pp. 62–70, 2004.
39. S. Zhang, G. H. Huff, J. Feng, and J. T. Bernhard, "A pattern reconfigurable microstrip parasitic array," *IEEE Transactions on Antennas and Propagation*, vol. 52, pp. 2773–2776, 2004.
40. E. Erdil, K. Topalli, M. Unlu, O. A. Civi, and T. Akin, "Frequency tunable microstrip patch antenna using RF MEMS technology," *IEEE Transactions on Antennas and Propagation*, vol. 55, no. 4, pp. 1193–1196, 2007.
41. R. Al-Dahleh, C. Shafai, and L. Shafai, "Frequency agile microstrip patch antenna using reconfigurable MEMS ground plane," *Microwave and Optical Technology Letters*, vol. 43, no. 1, pp. 64–67, 2004.
42. X. S. Yang, B. Z. Wang, and W. Wu, "Pattern reconfigurable patch antenna with two orthogonal quasi-Yagi arrays," in 2005 *IEEE Antennas and Propagation Society International Symposium*, Washington, DC, USA, vol. 2B, pp. 617–620, 2005.
43. C. W. Jung, M. J. Lee, G. P. Li, and F. De Flaviis, "Reconfigurable scan beam single-arm spiral antenna integrated with RF-MEMS switches," *IEEE Transactions on Antennas and Propagation*, vol. 54, no. 2, pp. 455–463, 2006.
44. T. Kawase, S. Moriya, C. J. Newsome, and T. Shimoda, "Inkjet printing of polymeric field-effect transistors and its applications," *Japanese Journal of Applied Physics*, vol. 44, p. 3649, 2005.
45. B. S. Cook, B. Tehrani, J. R. Cooper, and M. M. Tentzeris, "Multilayer inkjet printing of millimeter-wave proximity-fed patch arrays on flexible substrates," *IEEE Antennas and Wireless Propagation Letters*, vol. 12, pp. 1351–1354, 2013.
46. S. Ahmed, F. Tahir, A. Shamim, and H. M. Cheema, "A compact Kapton-based inkjet printed multiband antenna for flexible wireless devices," *IEEE Antennas and Wireless Propagation Letters*, vol. 14, pp. 1802–1805, 2015.

47. K. Hettak, T. N. Ross, R. James, A. Momciu, and J. S. Wight, "Flexible plastic substrate-based inkjet printed CPW resonators for 60 GHz ISM applications," in *Proceedings of the European Microwave Conference (EuMC)*, Rome, Italy, pp. 1194–1197, 2014.

48. L. Yang, A. Rida, R. Vyas, and M. M. Tentzeris, "RFID tag and RF structures on a paper substrate using inkjet-printing technology," *IEEE Transactions on Microwave Theory and Techniques*, vol. 55, no. 12, pp. 2894–2901, 2007.

49. G. Shaker, S. Safavi-Naeini, N. Sangary, and M. M. Tentzeris, "Inkjet printing of ultrawideband (UWB) antennas on paper-based substrates," *IEEE Antennas and Wireless Propagation Letters*, vol. 10, pp. 111–114, 2011.

50. L. Yang, R. Zhang, D. Staiculescu, C. P. Wong, and M. M. Tentzeris, "A novel conformal RFID-enabled module utilizing inkjet-printed antennas and carbon nanotubes for gas-detection applications," *IEEE Antennas and Wireless Propagation Letters*, vol. 8, pp. 653–656, 2009.

51. G. McKerricher, J. G. Perez, and A. Shamim, "Fully inkjet printed RF inductors and capacitors using polymer dielectric and silver conductive ink with through vias," *IEEE Transactions on Electron Devices*, vol. 62, no. 3, pp. 1002–1009, 2015.

52. S. N. Zabri, R. Cahill, G. Conway, and A. Schuchinsky, "Inkjet printing of resistively loaded FSS for microwave absorbers," *Electronics Letters*, vol. 51, no. 13, pp. 999–1001, 2015.

53. S. Choi, W. Su, M. M. Tentzeris, and S. Lim, "A novel fluid-reconfigurable advanced and delayed phase line using inkjet-printed microfluidic composite right/left-handed transmission line," *IEEE Microwave and Wireless Components Letters*, vol. 25, no. 2, pp. 142–144, 2015.

54. H. K. Kim, K. Ling, K. Kim, and S. Lim, "Flexible inkjet-printed metamaterial absorber for coating a cylindrical object," *Optics Express*, vol. 23, no. 5, pp. 5898–5906, 2015.

55. M. Walther, A. Ortner, H. Meier, U. Löffelmann, P. J. Smith, and J. G. Korvink, "Terahertz metamaterials fabricated by inkjet printing," *Applied Physics Letters*, vol. 95, p. 251107, 2009.

Chapter 6

Organic Fibers in Printed Electronics: Processing and Applications

Nada Omran,[a] **Ahmed H. Hassanin,**[b] **and Nader Shehata**[c]

[a]*Center of Smart Materials, Nanotechnology, and Photonics at Smart CI Research Center of Excellence, Faculty of Engineering, Alexandria University, Alexandria, Egypt*
[b]*Department of Textile Engineering, Faculty of Engineering, Alexandria University, Alexandria, Egypt*
[c]*Department of Engineering Mathematics and Physics, Alexandria University, Alexandria, Egypt*

eng-nada.hossam2018@alexu.edu.eg, ahassanin@alexu.edu.eg, nader.shehata@alexu.edu.eg

6.1 Introduction

The convergence of technical advancement and scientific discovery that resulted in the introduction of diodes, transistors, and ultimately compact chips in favor of giant vacuum tubes has contributed to the expansion of the electronics sector [1]. Printed electronics is the field of electronics manufacturing and material science, which refers to a method of producing electronic items such as electrical circuits, displays, sensors, and RFID tags, using organic and inorganic materials through different printing

Flexible Inkjet-Printed Antennas: Design, Fabrication, and Applications
Edited by Nader Shehata

ISBN 978-981-4968-87-4 (Hardcover), TBA (eBook)
www.jennystanford.com

DOI: 10.1201/9781003774754-6

technologies [2]. The field of printed electronics is quite broad with a goal of being able to print every component of the system at once, encompassing a wide range of materials and printing techniques creating discrete components, systems, or entire devices using sequential and/or parallel material additive technologies like screen printing, ink-jet printing, or direct writing [3].

A printed electronics device is created in its most 'pure' form by directly constructing conductive lines as well as all essential passive and active electronic components on (typically) organic, flexible substrates [2] such as glass, fabrics, paper, plastics, and metal, in addition to polymer substrates, which allow for innovative transparent and elastic electronics [4]. This is a revolutionary method that combines an existing manufacturing technology with innovative materials. It is expected to be widely employed in a variety of applications, although it is still in development and has yet to find a commercial niche [5].

Printed electronics began as a possible low-cost substitute for silicon-based electronics, but has now evolved into a complementing technology. Low-cost printed electronic components aimed at high-volume market segments where the high performance of conventional electronics is not necessary have created new possibilities thanks to relatively simple techniques that take less time and waste less material [6]. Another significant benefit of printed electronics is the ability to fabricate electronics in a flexible, ultra-thin form using a mix of specialty materials and low-cost, large-area fabrication procedures (such as printing) [7]. They are light, bendable, foldable/rollable, conformal, and wearable, and they allow for the creation of devices/sensors on a number of substrates, each shaped and customized to work differently [3].

Greater complexity and the integration of multiple components, materials, and processes are the main obstacles for smart systems. It is necessary to address higher definition, registering accuracy, and consistency over wide regions. Processing time, yield/uniformity monitoring, drying and curing in temperature/time tolerances, safety, health, and environmental effects are all issues that must be addressed in the manufacturing process [3].

Printing materials require improved processability, electrical performance, and environmental stability. Printability, chemical and physical compatibility of materials, as well as substrate qualities, including compressibility, smoothness, wettability, porosity, ink uptake, and ink properties (ink stability, viscosity, rheological properties, and so on), are some of the problems that must be overcome for large-scale manufacturing of printed electronics.

In this chapter, a thorough study of the advances in printing technology is introduced. Different techniques, along with a survey of the materials used in the final applications are presented. Nevertheless, the characterizations needed to qualify the produced substrates and the end-of-life options are also included, presenting a complete basic information source.

6.2 Printing Mechanisms

Electronics printing uses well-known printing methods from the graphic arts sector, each of which has advantages and disadvantages. Inkjet, screen printing, flexography, rotogravure, pad printing, offset, and embossing are among the printing methods utilized [8] with different characteristics for each method, as shown in Table 6.1. Sensors, sensor systems, batteries, capacitors, transistors, photovoltaic cells, memory, electroluminescent structures, and large-scale printings—screens or light panels—are all components of printed devices.

6.2.1 Screen Printing

The first method for printing electronics is screen printing. It is also the most affordable, straightforward, and adaptable method of printing. It is used to print electrodes for transistors, capacitors, and membrane switches. Among others, Janczak, Wróblewski, and Jakubowska [9] conducted studies on the screen printing of sensors on polymer substrates. They used this technique for the fabrication of graphene nanoplatelets and carbon nanotubes.

Table 6.1 Main characteristics of printing technologies for printed electronics [19, 20]

Printing technology	Offset	Flexography	Gravure	Screen	Inkjet
Ink viscosity (Pa s)	5–100 40–100	0.05–0.5	0.01–0.2 0.05–2	0.5–50	0.001–0.04
Layer thickness (μm)	0.5–2 0.5–1.5	0.04–2.5 1–2.5	<0.1–8 <0.1–5	0.015–100 Up to 12	0.05–20 0.5–15
Feature size (μm)	10–50	80	75	20–100	20–50
Registration (μm)	>10	<200	>20	>25	5–20
Throughout (m^2/s)	3–30	3–30	3–60	2–3	0.01–0.5
Resolution (lines/cm)	100–200	60	100	50	60–250
Printing speed (m/min)	200–800	100–500	100–1000	10–15	15–500

6.2.2 Pad Printing

Pad printing is yet another basic printing technique. In place of screen printing, it may be used to print transistor electrodes [10].

6.2.3 Inkjet Printing

CPW inductors may be manufactured using inkjet technology when using printed circuit boards (PCBs). This method was utilized by Menicanin et al. [11] to print using Ag nanoparticle inks on polyimide film (Kapton). Menicanin et al. [12] produced two- and three-layer coils in subsequent investigations. Its effectiveness was increased by using many layers.

The technology struggles to produce thick, 3D structured features since the inkjet printing ink selection is restricted to fluids with relatively low viscosities [13, 14].

6.2.4 Flexography

Flexography is typically used to print organic electrical circuits, such as OLED and photovoltaic cells (solar batteries), batteries, and RFID antennas. OLED diodes are used to make lighting (diode lamps) and diode displays (for TVs, computers, etc.). Intelligent labels that are printed on polyester film may be used to track the temperature of pharmaceuticals being transported (thin film) [15–18].

6.3 Materials Used

Electronic devices, flexible substrates, interconnects, bonding layers, and encapsulating layers are often used to build electronic systems. The uniformity of printed layers and the adhesion of layers to other layers are very important electrical properties of ink [21]. With the exception of the elastomeric substrates, flexible electronic systems are nearly identical to conventional electronic systems [22]. To produce a functional device, it is essential to define the printing procedure along with the right material such that the surface of the substrates would be compatible with the functional ink used. The production of printed electronics depends heavily on print quality, uniformity, smoothness, and patterning challenges, which include resolution, design guidelines, accuracy, registration, and yield [23].

A slight change in the structure or the material not only opens up a wide range of novel uses but also creates enormous difficulties in the manufacture of both materials and devices.

6.3.1 Ink Materials

Electronically active materials with conducting, semiconducting, luminous, electrophoretic, or electrochromic qualities are known as functional inks or materials for printed electronics [3, 24]. They can be evaporated or based in a solution, organic, or inorganic. Depending on the needs of the device application and the manufacturing process utilized, a particular material may be chosen in a given situation. To achieve outstanding printing quality, functional inks must have high dispersion ability and must have adequate fluidic qualities. To obtain a high ink transfer rate and construct the required pattern with high precision, the viscosity and wetting qualities must be properly tuned [25].

Printing inks are designed to provide pictures and a specific message and can be spot colors, created specifically for a given product, or they can incorporate functionality, such as security measures to prevent fraud, or they can include intelligence capabilities [3].

The pigments, solvents, resins, and/or polymers are the major ingredients of printing (graphic) inks, and in addition, in concentrations not exceeding 5%, give inks their final characteristics. To achieve the required electrical properties of conductivity, functional materials are used in place of pigments in functional inks, as shown in Table 6.2. Regarding their composition, inks may be conductive, semiconductive, dielectric, thermos-chromic, photochromic, magnetic, or may have another specific activity [3]. Due to their quick flow in electrical devices, some inks used in printed electronics contain nanoparticles formed of metal or ceramic, they are helpful in electronic, electrical, or optical devices because they are simple to operate and can be put together reversibly in one or more ways [26].

Table 6.2 Conductivity of different types of inks [3, 27]

Material	Conductivity
Silver nanoparticles for inkjet printing	40 MS/m
Copper nanoparticles for inkjet printing	23 MS/m; 8 MS/m; 18 MS/m
Silver nanoparticles for gravure or screen printing	50 MS/m
Silver particles for flexographic printing	1.7 MS/m
Silver particles for screen printing	1.3 MS/m; 3.9–7.8 MS/m
Graphene for gravure or screen printing	0.03 MS/m

6.3.1.1 Conductive inks

Metallic particles, either nanoparticles or micron-sized particles (flakes) that are extremely conductive, make up the main component of conductive inks. Metallic colors like silver, gold, copper, nickel, or platinum are frequently utilized [23]. They are used to print electrical devices such as printed heater circuits, additive circuits, flexible circuits, membrane switches, sensors, electroluminescent displays, and resistance pads.

Because of their poor solubility, stability, and processability, conductive organic polymers in solvents, metallic flakes suspended in binders, or organic-metallic mixes are more challenging to include in inks. Various optoelectronic devices, such as polymer light-emitting diodes and polymer transistor circuits, use conductive polymers and semiconducting inks [28, 29], which are made of semiconducting organic polymers in solvents, inorganic nanoparticles suspended in carrier fluids, or organic-inorganic blends [30]. Organic polymers in solvents, organic polymer thermosets, or organic polymers packed with ceramics make up dielectric inks.

6.3.1.2 Chromogenic inks

Materials that change their optical characteristics in response to an external input are called chromogenic [31]. A change in temperature (thermochromic materials), exposure to light (photochromic materials), a shift in the pH of the surrounding solution (chemochromic materials), the application of an electric potential (electrochromic materials), or other factors, such as radiation to the magnetic field, the presence of pathogens, radioactivity, or IR radiation [31, 32], could all cause this. Thermochromic and photochromic inks are the two main categories of color-changing inks.

6.3.1.3 Photochromic inks

The incoming light intensity or spectral composition affects how photochromic inks alter their color [32]. When not exposed to UV light, they are reversible and almost undetectable; nonetheless, UV light exposure triggers a quick transition and darkens the ink [3]. In contrast to organic photochromic chemicals, which are often unstable organic molecules like spirobenzopyrans, fulgides, chromenes, diarylethenes, and inorganic and biological photochromes are less effective overall.

6.3.1.4 Thermo-chromic inks

Special effects inks called thermos-chromic inks change color when exposed to certain temperatures. Since they are thermosensitive, they react to heat by changing color, whether from colorless to colorful, from colored to colorless, or from one color to another. Each color has a set temperature range, and when the necessary temperature is achieved (often referred to as the activation temperature or activation point), the color changes [3]

6.3.2 Substrate Materials

Materials of the substrates can be either stiff or flexible. To enable big areas, huge quantities, and a greater degree of design flexibility, flexible and possibly inexpensive substrates are sought [24]. The type of substrate has an influence on the quality of printed electronics [33], which is why there are a variety of materials and techniques that can be used to produce substrates for printed electronics.

The three most used substrate types are thin glass, plastic films, and metal foils [22]. None of them are faultless. Brittleness brought on by crack spread is the main issue with thin glass. The thermal mismatch stress, which restricts the processing temperature and the serving environment, and the simple entry of oxygen and water, which might damage the functional component over time, are the two main problems that plastic films face. The last category is metal foils, with stainless steel being the most popular, which often has poor optical transmittance and, more dangerously, the system can fail due to surface roughness.

While it is easier to print on paper rather than films [34], coated and uncoated paper substrates are also becoming more and more common as building blocks for printed electronics. Some studies also proved that recycled fibers and papers can be utilized as cardboard for usage as substrates for printed electronics [35, 36]. Also, cellulosic substrates are becoming more popular due to the higher efficiency of printed electronics especially when compared with PET substrates [37]. The introduction of affordable, flexible screens has increased interest in silicon.

Interestingly, it could be intriguing to deposit functional nanofibers on metal substrates in order to directly fabricate flexible devices. Nanofibers are used also in the drying step to enhance the uniformity of the printing ink after liquid evaporation and to avoid issues including the coffee-ring effect, the deterioration of paint coats, the clarity of printed characters on paper, and the conductivity of printed electronics that are caused by uneven drying [38].

6.4 Processing of Different Forms of Printed Electronics Substrates

The stability and lifespan of the entire system can be increased once the functional parts are mounted on rigid substrates because substrates merely serve as carriers for functional parts. The failure of electronics mostly results from the disability of those functional parts, which in most cases is caused by being damaged by overstretched substrates. That is why recent studies are working on developing a combination of hard and flexible electronics through different techniques and fabrication methods.

6.4.1 Flexible Hybrid Electronics

A large number of printed electronic parts and silicon integrated circuits are combined on a flexible substrate to create flexible hybrid electronics (FHE), to convert physical and chemical qualities like temperature, light, pressure, chemical concentrations, etc. into electrical impulses [39]. FHE is a flexible, effective, economical, and large-area-compliant electronics platform. It is both powerful and adaptable. FHE systems employ printed sensors and generally printing is frequently employed as a fabrication method. The electrical signals are cleaned up and processed using silicon integrated circuits (ICs), after which it is either shown on a printed display or broadcast to an external host using printed antennas. The system is often powered by an energy harvester or a printed battery.

6.4.2 Polymer Substrates

Polymer substrates provide new avenues and enable the development of transparent and elastic electronics. Polyimide (PI), polyamideimide (PAI), polycarbonate (PC), polyethylene naphthalene (PEN), polyethylene terephthalate (PET), polyethylene (PE), and polyarylethersulfone (PES) are the most popular plastic substrates (PE) [4].

The display in the work by Yakoh et al. [40] was created on PET film via screen printing, and successive layers were added. The display has been demonstrated to continue functioning when being bent. It is challenging to combine with garments due to the substrate of foil. PET film is not flexible enough to stretch or bend with extremely tiny radii. Since the antennas must be smaller, lighter, and less expensive than typical antennas manufactured by etching, printing the antennas on flexible substrates is very important [4]. Research on the utilization of silver nanowire pastes for printing antenna conductors on PET film using a low-temperature technique was done by Komoda et al. [37], which proved that for the production of printed electronics on polymer substrates, low-temperature techniques are appropriate.

6.4.3 Fabric

Stretchable electroluminescent (EL) structures were created by Janczak et al. [41] by utilizing newly discovered polymer compositions based on a variety of various nanomaterials, including dielectric, carbon nanotubes, conductive platelets, and luminophore nanopowders. Using the screen-printing technique, multi-layered EL structures have been printed directly on fabrics. The generated composite pastes' suitable rheological characteristics, as well as their applicability for printed electronics, have been proven. The new materials' mechanical strength and resistance to washing and ironing have been evaluated on the structure they have produced.

Technology that won't interfere with its functioning conditions is needed for the development of integrated cloth electronics [42–44]. For integrated devices to provide acceptable user comfort, they must be elastic and flexible. The device must be taken into consideration for potential damage from variables like washing or ironing.

Display screens are often created in wearables using a variety of printing techniques, including inkjet or screen printing [45, 46]. They are created by adding the proper functional substrate layers. An insulating dielectric layer, a luminophore, and two layers of conductive electrodes make up the display. Typically, composites of dissolved elastic polymer and filler powders are used to print the layers. This method frequently calls for high voltage, typically between 100 V and 1000 V, as well as a variety of frequencies [47–49]. In order to assure user safety, insulating layers should be added to the application.

Verboven et al. [50] used a screen-printing technique on a polyester cloth to generate event and brilliant electroluminescence. Both investigations employed polyester cloth rather than cotton, which rendered it unsuitable for wearable electronics and daily usage.

An electroluminescent display was powered by friction between specifically prepared and woven threads in a self-powered motion-driven triboelectric electroluminescence system, which was used in many ways [51]. Such displays are simple to put into garments, but because of their extremely low luminescent intensities, they are not visible in strong light.

Słoma et al. [48] made attempts to create the display on textiles by printing layers of the display directly on cotton fabric. The printing inks were based on a poly(methyl methacrylate) (PMMA) carrier, which was employed for the deposition. While the display's bending caused uneven illumination and, ultimately, actual damage.

6.4.4 Nanofibers

Elastomeric substrates replace their rigid counterparts in the realm of flexible electronics, enabling intriguing applications including solar cells, electronic skins, and flexible displays. Nanoscale dimensions result in unique material characteristics, as is the case with nanofibers that have a high surface-to-volume ratio and a big aspect ratio. Researchers are working to incorporate flexible electronics with nanofibers made of polymers, metals, carbon materials, composites, and even ceramics [22].

There are several well-established techniques for creating nanofibers, but each has advantages and disadvantages. Drawing [52], template synthesis [53, 54], self-assembly [55–58], phase separation [59–62], electrospinning [63–65], melt spinning [66–68], and solution blow spinning [69–71] are a few techniques that may be used to create pure nanofibers or metal, carbon, and ceramic nanofibers. A brief introduction to the most utilized nanofiber fabrication techniques is presented in this section.

Nanofiber patterns and structures have only been successfully fabricated on 2D surfaces [72–75]. To get compatibility with conformal electronics and 3D living creatures [76] and to give novel functions that only appear on multidimensional surfaces, functional nanofiber production on 3D surfaces is crucial [77, 78]. Through the use of the self-aligning nanojet (SA-N), Shin et al. demonstrated the direct printing of functional nanofiber on 3D surfaces [79].

With the capacity of small molecules to form self-assembled structures in very concentrated solutions or pure melts, which is dependent on the presence of sufficient intramolecular interactions among them, they can be directly spun into fibers. The most often reported small compounds [80–84] are amphiphiles and derivatives of cyclodextrin. For the purpose of creating nanofibers, a colloid with a dispersed phase (the suspended particles) and a continuous phase (the medium of suspension) has also been used [85]. In addition to two-dimensional (2D) nanostructures like clay tablets and graphene nanosheets, as well as one-dimensional (1D) nanostructures like Au nano-rods, carbon nanotubes, and Au or Ag nanowires [86], nanoparticles made of Au, Ag, and TiO_2, as well as particles made of metalorganic framework (MOF) compounds, are widely used as nanoscale components [87]. Different methods of treatment can be applied after the deposition of nanofibers to produce a variety of membranes. Nanofibers of ceramics, carbon materials, and activated carbon are often produced by applying post-treatments including calcination, carbonation, and activation after the nanofibers have been deposited [22].

6.4.4.1 Electrospinning

The method of electrospinning is fairly easy. The fundamental difficulty in electrospinning is maximizing many parameters to produce the desired nanofibrous membranes. It is a flexible method that creates extremely tiny threads of a polymer solution using electrostatic forces. Ranges from submicron to nanometers in diameter. Natural, synthetic, biodegradable, and non-degradable polymers, as well as their mixtures, can all be employed as polymer materials. To introduce charges into a polymer solution, a high voltage is used. A droplet at the needle's tip forms a Tylor cone when the charges reach a critical point, and a fluid jet shoots from it in the direction of the grounded collector, which has a lower potential [63–65, 88, 89].

Near-field electrospinning (NFES) was created in order to achieve the direct printing capability of nanofiber by considerably lowering the applied voltage and tip-to-collector distance (TTCD) [72]. Due to the jet's lack of bending instability and the short TTCD, the jet may be deposited on a moving substrate in a controlled way. Through the mechanical drawing approach, continuous deposition of nanofiber across a sizable area was demonstrated using NFES [73]. Low-voltage NFES has been used on nonplanar surfaces with a fiber diameter as small as 16 nm by further lowering the applied voltage to roughly 200 V [90].

By synchronizing the jet impact speed and collector speed, perfect printing of a continuous nanofiber for complicated designs on a microscale was achieved in addition to simple nanofiber lines [74]. The direct-printing capability of NFES has facilitated the use of functional nanofiber in a variety of fields, such as electronics [91], optoelectronics [92], sensors [93], and energy harvesting [75]. As said, since its initial presentation in 2006, NFES has made great progress as one of the direct-printing techniques. Concentric electrospinning is another form of electrospun nanofibers that was used for flexible electronics to create a core-shell nanostructure by using a concentric nozzle with an inner and an outer nozzle [94, 95].

6.4.4.2 Solution blow spinning

In comparison to other methods of producing fibers, solution-blown spinning (SBS) can manufacture nanofiber mats for large-scale production [70], and it is becoming more popular because of its straightforward operation and high output [69]. In addition to planer and non-planer surfaces, this also opens up the option of creating nanofibers on a variety of substrates for direct application in intended applications [96]. The capacity of SBS to achieve deposition rates around 10 times greater than those of traditional electrospinning [97–99] is its major characteristic. Pressurized gas and polymer solution are simply provided inside the same device (i.e., nozzle). An atomization machine with an internal mixing capability can be utilized in conjunction with a commercial airbrush (gravity-fed brush) [100, 101].

6.4.4.3 Self-assembly

Self-assembly is the process of creating nanofibers out of tiny molecules as building blocks [55, 102–105]. The macromolecular nanofiber structure is determined by the same intermolecular forces that pack the smaller units tightly. The extension perpendicular to the plane, created when a tiny molecule is positioned in a concentric manner and the bonds are formed concentrically, creates the longitudinal axis of the nanofiber.

6.4.4.4 Phase separation

The fundamental tenet of this approach is the separation of phases due to physical incompatibility, followed by the removal of one phase and the retention of the other. After being dissolved in a solvent, a polymer goes through gelation [106]. Five simple steps—polymer dissolving, gelation, solvent extraction, freezing, and freeze-drying—can be used to sum up this procedure.

6.4.5 3D Printing

A new technique called 3D printing of functional materials and devices has the potential to provide manufacturers of electronic devices more flexibility in terms of material choice, 3D device form factor, morphology of target surfaces, and autonomy.

Light-based and ink-based printing are the two basic categories into which 3D printing can be divided [107]. Light-based printing typically uses concentrated laser beams to harden polymeric inks layer by layer while following a predetermined geometric form. Usually, photoinitiators are added to polymeric molecules before laser irradiation to hasten the cross-linking process.

On the other hand, ink-based 3D printing can be categorized as either extrusion-based or inkjet-based depending on the method of ink deposition. High-frequency inkjet printing disperses inks in the form of liquid droplets onto substrates after being triggered by heat or sonic means [14]. Inks such as liquid droplets, viscous polymers, or molten plastic filaments are extruded from printer nozzles when using ink-extrusion-based 3D printing (often with solenoid valves to control airflow). Industrial robots are equipped with high-precision computer numerical control (CNC) systems, which allow 3D printers adapted from these robotic systems to theoretically perform motions with resolutions as low as tens of nanometers. Direct ink writing, another name for extrusion-based 3D printing, allows active structures and gadgets to be printed on curved and flat surfaces. Due to the inks' pneumatic dispensing, their viscosities can vary greatly (up to 106 mPa s) and can have a variety of different characteristics, such as solutions, colloids, liquid metals, and polymeric resins [107, 108]. A new patent [109] describes a method for 3D printing active electronics using both conducting and semiconducting materials.

An attractive promise for going beyond planar and stiff traditional electronics is the capacity to directly 3D print electronics into the human body utilizing multipurpose inks. In this case, 3D printing offers three benefits [110]: (1) 3D printing enables patterning of electronic materials over multiscale and multidimensional surfaces that are characteristics of biology and different from planar conventional electronics; (2) nanoscale inks can be produced exhibiting high performance and functionality but lack the mechanical rigidity and brittleness of planar wafers such as Si, allowing functional inks to be interwoven with soft 3D printed hydrogels and polymers; and (3) these functional inks may be produced at high temperatures to create high-performance nanocrystalline materials, but can then be assembled afterward using an extrusion-based 3D printing method in ambient settings.

Wearable electronics [111], bionic skins [112], and on-body energy harvesting [113] are only a few examples of impact zones that might profit from the close and flexible contact that 3D printing enables between human tissue and electronics. To date, it has been shown that touch sensors and inductive coils may be 3D printed directly onto human skin and hand models, respectively [114, 115].

6.5 Applications

A variety of applications are now depending on printed electronics to convoy the advances in the market. Aside from passive ID cards printed on paper used for tickets or toys [3], one of the earliest uses of printed electronics, according to Kleper [116], is RFID embedded in boxes. Roll-to-roll manufactured flexible lithium polymer batteries, printed electrodes for glucose test strips, printed RF-driven smart objects, printed antennas for RFID tags, and large-area organic pressure sensors for retail logistics are other applications. All these applications are just a few of the first commercial applications that have been available since 2010 [3]. In the electronics manufacturing industry, printed passive components based on printable conductors and dielectrics, such as RFID antennas, are already widely known. Roll-to-roll printing has been used to create printed electrochromic displays, sensors, and wireless power transfer devices [117].

Integrated solutions are already beginning to appear on the market, including antitheft/forgery labels, temperature sensors, and smart labels. Printed batteries, printed memory, active (transistors, diodes, logic circuits, display elements), passive (resistors, capacitors, conductors), and active (transistors, diodes, logic circuits, display elements) devices are just a few examples of the various components that make up integrated smart systems. These components may also be assembled from a variety of independently manufactured components. In this section, we present a brief example of some of the most known applications for printed electronics.

6.5.1 Solar Cells

Solar cells are created using an inkjet and screen technique. The finger grid, bus line, and the black plane contact are rotogravure printed using Ag pastes. Inkjet lines (also known as doping lines) are frequently formed underneath Ag lines on the outside surface [15, 118–121].

6.5.2 Electrochromic Devices

Numerous suggested optical security devices with characteristics that are optically changeable when a voltage is applied have also been made possible by the printed electronics industry's rapid development. Due to their low switching potential and memory effect, electrochromic materials are frequently chosen in this situation as well (colored state is maintained once voltage is removed). Hewlett-Packard, for instance, recommended using an electrochromic device for security tags [122]. Another electrochemical-based illustration for security papers like passports and smart cards was proposed [123]. In this instance, the RFID technology utilized in the passport's electronic components might also be used to transmit an RF signal to power the feature.

The entire thickness was based on a five-to-seven-layer structure and was around 30 m. Their usage as banknotes may be constrained by their excessive thickness. This technology's foundation was a white reflector with an electrochromic viologen matrix made of TiO_2 attached to it [124]. Others have created methods to decrease the overall number of layers as well as the driving forces that might lower costs and enable simple authentication [125]. Electrochromic paper has also been proposed as a viable option for banknotes, tokens, and coupons [126]. All of these examples provide a novel method of authentication, but they are primarily intended to be used as a second-level authentication feature for knowledgeable staff rather than for the general public.

6.5.3 Radio Frequency Identification (RFID)

The RFID system, in which the circuits and batteries in the antennas may be printed, is the most widely used example of printed electronics. These technologies make it possible to detect, classify, and manage the product flow. Since the antennas must be smaller, lighter, and less expensive than typical antennas manufactured by etching, printing them on flexible substrates is particularly important [4].

6.5.4 Thin Film Transistor (TFTC) Tags

The goal is to print the printed TFTC electrical equivalent of a chip on a regular tag, albeit with substantially lesser performance. For instance, it is currently challenging or even impossible to identify TFTC tags using the same readers and protocols as the traditional RFID technology. The "chip" that was created is actually rather basic because it lacks a microprocessor and reprogrammable memory. A few bits make up the memory size. The tag support has transistors that are directly printed on it. Completely printed RFID tags that operate in the HF band at 13.56 MHz have been presented since 2007 by businesses like PolyIC and Kovio [127].

6.5.5 Food Packaging

Advanced packaging improves on the fundamental capabilities of conventional food packaging. Modern food packaging aims to improve food quality, product safety, shelf life, and informational content. Several words, including smart, active, intelligent, interactive, clever, responsive, and diagnostic, are used to characterize packaging with additional features. While active packaging (AP) provides food with improved protection, intelligent packaging (IP) is the component in charge of monitoring the environment within or around packaging and responding accordingly. The advantages of active and intelligent technologies are combined in smart packaging or a comprehensive package concept. The goal of IP is to monitor and indicate if food is still fresh [3].

Data carriers, such as barcodes and RFID tags, which are specially designed for storage, distribution, and traceability functions, are included in the package to create an IP system. Along with sensors, which enable a rapid and accurate measurement of the compounds in meals. In addition to indications that seek to increase convenience for customers and/or educate them about the quality of food. They differ from one another in terms of their physical makeup, the volume and kind of data that can be sent, and the methods used to collect and transmit the data.

6.5.6 End-of-Life (EoL) Options

When environmental, safety, and health concerns are taken into account, new technologies, like printed electronics, are being developed, and the benefits are maximized. It is important to think about what happens to the items after usage today. Future trash will increase as there are more items available. Can printed electronic applications be recycled and reused, and do they present any potential threats to the environment, safety, and health if they wind up in a landfill?

Since the lifespan of electronic applications is often rather brief, it is commonly believed that this phase is of major relevance. Wastes from electrical and electronic equipment are the waste category with the quickest rate of growth [5]. Older versions are too expensive to maintain, and they quickly become out of date. A short lifetime indicates relatively high product use. Due to the numerous potential uses for printed electronics, there will be more waste in the future. The usage of innovative materials is one of the primary environmental issues with printed electronics. In the usage phase of nanotechnology-enabled devices, exposure to nanoparticles is extremely improbable, but during the EoL phase of the applications, environmental concerns might develop, claims a paper from Finland's Environmental Administration.

References

1. Mahajan, G., M. Kaur, and R. Gupta, Green functionalized nanomaterials: Fundamentals and future opportunities, in *Green Functionalized*

Nanomaterials for Environmental Applications. 2022, Elsevier. pp. 21–41.

2. Swenson, O. and V. Marinov, Laser processing of direct-write Nano-sized materials, in *Advances in Laser Materials Processing*. 2010, Elsevier. pp. 671–694.
3. Gregor-Svetec, D., Intelligent packaging, in *Nanomaterials for Food Packaging*. 2018, Elsevier. pp. 203–247.
4. Izdebska, J. and Thomas, S., Eds, Applications of printed materials. 2016, pp. 371–388.
5. Keskinen, M., End-of-life options for printed electronics, in *Waste Electrical and Electronic Equipment (WEEE) Handbook*. 2012, Elsevier. pp. 352–364.
6. Kantola, V., et al., Printed electronics, now and future. 2009. 63: p. 204.
7. Lupo, D., et al., OE-A roadmap for organic and printed electronics, in *Applications of Organic and Printed Electronics*. 2013, Springer. p. 1–26.
8. Prudenziati, M. and J. Hormadaly, *Printed Films: Materials Science and Applications in Sensors, Electronics and Photonics*. 2012, Elsevier.
9. Janczak, D., et al., Screen printed resistive pressure sensors fabricated from polymer composites with carbon nanotubes. 2012. 3(4): pp. 14–19.
10. Knobloch, A., A. Bernds, and W. Clemens. Printed polymer transistors, in *First International IEEE Conference on Polymers and Adhesives in Microelectronics and Photonics. Incorporating POLY, PEP & Adhesives in Electronics*. Proceedings (Cat. No. 01TH8592). 2001. IEEE.
11. Menicanin, A., et al., Ink-jet printed CPW inductors in flexible technology, in *2012 Proceedings of the 35th International Convention MIPRO*. 2012. IEEE.
12. Menicanin, A. B., et al., Improved performance of multilayer CPW inductors on flexible substrate. 2014. 50(11): pp. 1–4.
13. Calvert, P. D., Inkjet printing for materials and devices. 2001. 13(10): pp. 3299–3305.
14. Zhang, Y., et al., Printing, folding and assembly methods for forming 3D mesostructures in advanced materials. 2017. 2(4): pp. 1–17.
15. Suganuma, K., *Introduction to Printed Electronics*. Vol. 74. 2014, Springer Science & Business Media.

16. Besganz, A., et al., Inkjet printing as a flexible technology for the deposition of thermoelectric composite structures. 2014. 15: pp. 99–106.

17. Sloma, M., et al., Graphene applications with printed electronics technology, in *2013 European Microelectronics Packaging Conference (EMPC)*. 2013. IEEE.

18. Mäkelä, T., Towards printed electronic devices. 2008.

19. Nomikos, S., et al., Printed technologies for intelligent packaging applications and their impact on printed electronics market. 2014. 1(1): pp. 1–4.

20. Rosa, P., Minimal computation structures for visual information applications based on printed electronics. 2015.

21. Joshi, S. S., Evaluation of silver/graphite ink blends for use in printed electronics. 2011.

22. Liao, S., Y. Huang, and H. Wu, Functional nanofibers for flexible electronics, in *Smart Sensors and Systems*. 2017, Springer. pp. 335–358.

23. Aijazi, A. T., Printing functional electronic circuits and components. 2014: Western Michigan University.

24. Kirchmeyer, S., The OE-A roadmap for organic and printed electronics: Creating a guidepost to complex interlinked technologies, applications and markets. 2016. 3(1): p. 010301.

25. Wang, Z., et al., Development of MnO_2 cathode inks for flexographically printed rechargeable zinc-based battery. 2014. 268: pp. 246–254.

26. Kosmala, A., et al., Synthesis of silver nano particles and fabrication of aqueous Ag inks for inkjet printing. 2011. 129(3): pp. 1075–1080.

27. Nilsson, H.-E., et al., System integration of electronic functions in smart packaging applications. 2012. 2(10): pp. 1723–1734.

28. Villani, F., et al., Inkjet printed polymer layer on flexible substrate for OLED applications. 2009. 113(30): pp. 13398–13402.

29. Paul, K. E., et al., Additive jet printing of polymer thin-film transistors. 2003. 83(10): pp. 2070–2072.

30. Pekarovicova, A. and V. Husovska, *Printing Ink Formulations, Printing on Polymers*. 2016, New York: Elsevier.

31. Seeboth, A. and D. Lötzsch, *Thermochromic and Thermotropic Materials*. 2013, CRC Press.

32. Ferrara, M. and M. Bengisu, Materials that change color, in *Materials that Change Color*. 2014, Springer. pp. 9–60.

33. Salmerón, J. F., et al., Properties and printability of inkjet and screen-printed silver patterns for RFID antennas. 2014. 43(2): pp. 604–617.

34. Robinson, N. D. and M. Berggren, Printing organic electronics on flexible substrates. 2007.

35. Kavčič, U., et al., UHF RFID tags with printed antennas on recycled papers and cardboards. 2014. 48(2): pp. 261–267.

36. Gregor-Svetec, D., M. Pivar, and D. Muck, Properties of recycled papers important for printable antennas. 2015. 49(7–8): pp. 701–708.

37. Komoda, N., et al., Silver nanowire antenna printed on polymer and paper substrates, in *2012 12th IEEE International Conference on Nanotechnology (IEEE-NANO)*. 2012. IEEE.

38. Ooi, Y., et al., Suppressing the coffee-ring effect of colloidal droplets by dispersed cellulose nanofibers. 2017. 18(1): pp. 316–324.

39. Khan, Y., et al., A new frontier of printed electronics: Flexible hybrid electronics. 2020. 32(15): p. 1905279.

40. Yakoh, A., et al., Screen-printed electroluminescent lamp modified with graphene oxide as a sensing device. 2018. 10(24): pp. 20775–20782.

41. Janczak, D., et al., Stretchable and washable electroluminescent display screen-printed on textile. 2019. 9(9): p. 1276.

42. Kim, S. J., et al., Materials for flexible, stretchable electronics: Graphene and 2D materials. 2015. 45: pp. 63–84.

43. Rogers, J. A., T. Someya, and Y. Huang, Materials and mechanics for stretchable electronics. 2010. 327(5973): pp. 1603–1607.

44. Wu, H., et al., Energy harvesters for wearable and stretchable electronics: From flexibility to stretchability. 2016. 28(45): pp. 9881–9919.

45. de Vos, M., R. Torah, and J. Tudor. A novel pneumatic dispenser fabrication technique for digitally printing electroluminescent lamps on fabric, in *2015 Symposium on Design, Test, Integration and Packaging of MEMS/MOEMS (DTIP)*. 2015. IEEE.

46. de Vos, M., et al., Functional electronic screen-printing–electroluminescent lamps on fabric. 2014. 87: pp. 1513–1516.

47. Bredol, M. and H. S. Dieckhoff, Materials for powder-based AC-electroluminescence. 2010. 3(2): pp. 1353–1374.

48. Sloma, M., et al., Electroluminescent structures printed on paper and textile elastic substrates. 2014.

49. Ferri, J., et al., Integration of a 2D touch sensor with an electro-luminescent display by using a screen-printing technology on textile substrate. 2018. 18(10): p. 3313.

50. Verboven, I., et al., Printing smart designs of light emitting devices with maintained textile properties. 2018. 11(2): p. 290.

51. Park, H.-J., et al., Self-powered motion-driven triboelectric electro-luminescence textile system. 2019. 11(5): pp. 5200–5207.

52. Alghoraibi, I. and S. Alomari, Different methods for nanofiber design and fabrication, in *Handbook of Nanofibers*. 2018, Springer. pp. 1–46.

53. Long, Y., et al., Electrical properties of isolated poly (3, 4-ethylenedioxythiophene) nanowires prepared by template synthesis. 2009. 20(6): pp. 541–544.

54. Pan, L., et al., Conducting polymer nanostructures: Template synthesis and applications in energy storage. 2010. 11(7): pp. 2636–2657.

55. Hartgerink, J. D., E. Beniash, and S. I. Stupp, Self-assembly and mineralization of peptide-amphiphile nanofibers. 2001. 294(5547): pp. 1684–1688.

56. Lee, C. C., et al., Preparation and characterization of helical self-assembled nanofibers. 2009. 38(3): pp. 671–683.

57. Hartgerink, J. D., E. Beniash, and S. I. Stupp, Peptide-amphiphile nanofibers: A versatile scaffold for the preparation of self-assembling materials. 2002. 99(8): pp. 5133–5138.

58. Fu, I. W. and H. D. Nguyen, Sequence-dependent structural stability of self-assembled cylindrical nanofibers by peptide amphiphiles. 2015. 16(7): pp. 2209–2219.

59. Ichimori, T., et al., Morphological diversity and nanofiber networks of poly (p-oxybenzoyl) generated by phase separation during copolymerization. 2013. 128(2): pp. 1282–1290.

60. Shao, J., et al., Early stage evolution of structure and nanoscale property of nanofibers in thermally induced phase separation process. 2012. 72(10): pp. 765–772.

61. Sun, L. L., et al., Dependence of dielectric properties and percolative behavior on phase separation structure induced by heterogeneous carbon nanofiber distribution in polymer blend nanocomposites. 2011. 296(11): pp. 992–1001.

62. Zhao, J., et al., Structure, morphology and cell affinity of poly(L-lactide) films surface-functionalized with chitosan nanofibers via a solid–liquid phase separation technique. 2013. 33(3): pp. 1546–1553.

63. Begum, H. A. and M. K. R. Khan, Study on the various types of needle-based and needleless electrospinning system for nanofiber production. 2017. 6: pp. 110–117.
64. Bhardwaj, N. and S. C. Kundu, Electrospinning: A fascinating fiber fabrication technique. 2010. 28(3): pp. 325–347.
65. Bognitzki, M., et al., Nanostructured fibers via electrospinning. 2001. 13(1): pp. 70–72.
66. Balogh, A., et al., Melt-blown and electrospun drug-loaded polymer fiber mats for dissolution enhancement: A comparative study. 2015. 104(5): pp. 1767–1776.
67. Ellison, C. J., et al., Melt blown nanofibers: Fiber diameter distributions and onset of fiber breakup. 2007. 48(11): pp. 3306–3316.
68. Hassan, M. A., et al., Fabrication of nanofiber meltblown membranes and their filtration properties. 2013. 427: pp. 336–344.
69. Atif, R., et al., Study of air pressure and velocity for solution blow spinning of polyvinylidene fluoride nanofibres. 2021. 9(6): p. 1014.
70. Elnabawy, E., et al., Solution blow spinning of piezoelectric nanofiber mat for detecting mechanical and acoustic signals. 2021: p. 51322.
71. Gao, Y., et al., Recent progress and challenges in solution blow spinning. 2021. 8(2): pp. 426–446.
72. Sun, D., et al., Near-field electrospinning. 2006. 6(4): pp. 839–842.
73. Chang, C., K. Limkrailassiri, and L. Lin, Continuous near-field electrospinning for large area deposition of orderly nanofiber patterns. 2008. 93(12): p. 123111.
74. Shin, D., J. Kim, and J. Chang, Experimental study on jet impact speed in near-field electrospinning for precise patterning of nanofiber. 2018. 36: pp. 231–237.
75. Chang, C., et al., Direct-write piezoelectric polymeric nanogenerator with high energy conversion efficiency. 2010. 10(2): pp. 726–731.
76. Huang, Y., et al., Assembly and applications of 3D conformal electronics on curvilinear surfaces. 2019. 6(4): pp. 642–683.
77. Adams, J. J., et al., Conformal printing of electrically small antennas on three-dimensional surfaces. 2011. 23(11): pp. 1335–1340.
78. Ko, H. C., et al., A hemispherical electronic eye camera based on compressible silicon optoelectronics. 2008. 454(7205): pp. 748–753.
79. Shin, D., et al., Direct-printing of functional nanofibers on 3D surfaces using self-aligning nanojet in near-field electrospinning. 2020. 5(6): p. 2000232.

80. Xu, J. F., et al., Photoresponsive hydrogen-bonded supramolecular polymers based on a stiff stilbene unit. 2013. 125(37): pp. 9920–9924.

81. Nuansing, W., et al., Electrospinning of tetraphenylporphyrin compounds into wires. 2014. 31(1): pp. 88–93.

82. McKee, M. G., et al., Phospholipid nonwoven electrospun membranes. 2006. 311(5759): pp. 353–355.

83. Hunley, M. T., M. G. McKee, and T. E. Long, Submicron functional fibrous scaffolds based on electrospun phospholipids. 2007. 17(7): pp. 605–608.

84. Jørgensen, L., K. Qvortrup, and I. S. Chronakis, Phospholipid electrospun nanofibers: Effect of solvents and co-axial processing on morphology and fiber diameter. 2015. 5(66): pp. 53644–53652.

85. Xue, J., et al., Electrospinning and electrospun nanofibers: Methods, materials, and applications. 2019. 119(8): pp. 5298–5415.

86. Zhang, C.-L. and S.-H. Yu, Nanoparticles meet electrospinning: Recent advances and future prospects. 2014. 43(13): pp. 4423–4448.

87. Efome, J. E., et al., Insight studies on metal-organic framework nanofibrous membrane adsorption and activation for heavy metal ions removal from aqueous solution. 2018. 10(22): pp. 18619–18629.

88. Doshi, J. and D. H. Reneker, Electrospinning process and applications of electrospun fibers. 1995. 35(2–3): pp. 151–160.

89. Greiner, A. and J. H. Wendorff, Electrospinning: A fascinating method for the preparation of ultrathin fibers. 2007. 46(30): pp. 5670–5703.

90. Bisht, G. S., et al., Controlled continuous patterning of polymeric nanofibers on three-dimensional substrates using low-voltage near-field electrospinning. 2011. 11(4): pp. 1831–1837.

91. Chang, J., et al., Direct-write complementary graphene field effect transistors and junctions via near-field electrospinning. 2014. 10(10): pp. 1920–1925.

92. Di Camillo, D., et al., Near-field electrospinning of light-emitting conjugated polymer nanofibers. 2013. 5(23): pp. 11637–11642.

93. Liu, X., et al., All-printable band-edge modulated ZnO nanowire photodetectors with ultra-high detectivity. 2014. 5(1): pp. 1–9.

94. Tiwari, M. K., A. L. Yarin, and C. M. Megaridis, Electrospun fibrous nanocomposites as permeable, flexible strain sensors. 2008. 103(4): p. 044305.

95. Dhevi, D. M., et al., Electrospun PVDF-TrFE-based piezoelectric sensors, in *Advances in Polymer Materials and Technology*. 2016, CRC Press. pp. 241–270.
96. Daristotle, J. L., et al., A review of the fundamental principles and applications of solution blow spinning. 2016. 8(51): pp. 34951–34963.
97. Medeiros, E. S., et al., Solution blow spinning: A new method to produce micro-and nanofibers from polymer solutions. 2009. 113(4): pp. 2322–2330.
98. Oliveira, J. E., et al., Structural and morphological characterization of micro and nanofibers produced by electrospinning and solution blow spinning: A comparative study. 2013. 2013.
99. Wendorff, J. H., S. Agarwal, and A. Greiner, *Electrospinning: Materials, Processing, and Applications*. 2012, John Wiley & Sons.
100. Vural, M., et al., Sprayable elastic conductors based on block copolymer silver nanoparticle composites. 2015. 9(1): pp. 336–344.
101. Tutak, W., et al., The support of bone marrow stromal cell differentiation by airbrushed nanofiber scaffolds. 2013. 34(10): pp. 2389–2398.
102. Liu, G., L. Qiao, and A. J. M. Guo, Diblock copolymer nanofibers. 1996. 29(16): pp. 5508–5510.
103. Liu, G., et al., Polystyrene-block-poly (2-cinnamoylethyl methacrylate) nanofibers—preparation, characterization, and liquid crystalline properties. 1999. 5(9): pp. 2740–2749.
104. Yan, X., et al., Superparamagnetic triblock copolymer/Fe_2O_3 hybrid nanofibers. 2001. 113(19): pp. 3705–3708.
105. De Moel, K., et al., Polymeric nanofibers prepared from self-organized supramolecules. 2001. 13(12): pp. 4580–4583.
106. Ma, P. X. and R. Zhang, Synthetic nano-scale fibrous extracellular matrix. 1999. 46(1): pp. 60–72.
107. Raney, J. R. and J. A. Lewis, Printing mesoscale architectures. 2015. 40(11): pp. 943–950.
108. Kong, Y. L., et al., 3D printed quantum dot light-emitting diodes. 2014. 14(12): pp. 7017–7023.
109. McAlpine, M. C. and Y. L. Kong, 3D printed active electronic materials and devices. 2018, Google Patents.
110. Su, R., et al., 3D printed electronic materials and devices, in *Robotic Systems and Autonomous Platforms*. 2019, Elsevier. pp. 309–334.

111. Stoppa, M. and A. J. Chiolerio, Wearable electronics and smart textiles: A critical review. 2014. 14(7): pp. 11957–11992.

112. Hou, C., et al., Highly conductive, flexible, and compressible all-graphene passive electronic skin for sensing human touch. 2014. 26(29): pp. 5018–5024.

113. Mitcheson, P. D., et al., Energy harvesting from human and machine motion for wireless electronic devices. 2008. 96(9): pp. 1457–1486.

114. Guo, S. Z., et al., 3D printed stretchable tactile sensors. 2017. 29(27): p. 1701218.

115. Zhu, Z., et al., 3D printed functional and biological materials on moving freeform surfaces. 2018. 30(23): p. 1707495.

116. Kleper, M. L., Printed electronics and the automatic identification of objects. 2004.

117. Kang, H., et al., Fully roll-to-roll gravure printable wireless (13.56 MHz) sensor-signage tags for smart packaging. 2014. 4(1): pp. 1–7.

118. Liang, J., et al., Elastic dynamic analysis on the scraping paste mechanism of the printing equipment for silicon solar cells, in *International Conference on Intelligent Robotics and Applications*. 2014. Springer.

119. Liu, H.-C., et al., Inkjet printing for silicon solar cells. 2009. 9(5): pp. 495–498.

120. Pagliaro, M., et al., Flexible solar cells. 2008. 1(11): pp. 880–891.

121. Bahrami, A., et al., Photovoltaic cells technology: Principles and recent developments. 2013. 45(2): pp. 161–197.

122. Champion, D. A. and C. S. Betrabet, Security device. 2009, Google Patents.

123. Baloukas, B. and L. Martinu, Optical coatings for security and authentication devices, in *Optical Thin Films and Coatings*. 2018, Elsevier. pp. 633–666.

124. Corr, D., et al., Coloured electrochromic “paper-quality” displays based on modified mesoporous electrodes. 2003. 165(1–4): pp. 315–321.

125. Rougier, A., et al., Electrochromism: From oxide thin films to devices, in *Oxide-Based Materials and Devices VI*. 2015. SPIE.

126. Monk, P. M., F. Delage, and S. M. C. Vieira, Electrochromic paper: Utility of electrochromes incorporated in paper. 2001. 46(13–14): pp. 2195–2202.

127. Vena, A., E. Perret, and S. Tedjini, The latest developments on chipless RFID technologies, in *Chipless RFID Based RF Encoding Part*. 2016, Elsevier. pp. 27–62.

Index

For Product Safety Concerns and Information please contact our EU representative GPSR@taylorandfrancis.com Taylor & Francis Verlag GmbH, Kaufingerstraße 24, 80331 München, Germany